Microstructure of
METALS
AND
ALLOYS

An Atlas of Transmission
Electron Microscopy Images

Microstructure of
METALS
AND
ALLOYS

An Atlas of Transmission
Electron Microscopy Images

Ganka Zlateva
Zlatanka Martinova

CRC Press
Taylor & Francis Group
Boca Raton London New York

CRC Press is an imprint of the
Taylor & Francis Group, an **informa** business

CRC Press
Taylor & Francis Group
6000 Broken Sound Parkway NW, Suite 300
Boca Raton, FL 33487-2742

First issued in paperback 2019

ISBN-13: 978-1-4200-7556-4 (hbk)
ISBN-13: 978-0-367-38734-1 (pbk)

Library of Congress Cataloging-in-Publication Data

Zlateva, Ganka.
    Microstructure of metals and alloys : an atlas of transmission electron microscopy images / Ganka Zlateva and Zlatanka Martinova.
       p. cm.
    Includes bibliographical references and index.
    ISBN 978-1-4200-7556-4 (alk. paper)
    1. Metals. 2. Alloys. 3. Microstructure. I. Martinova, Zlatanka. II. Title.

TN690.Z4585 2008
669'.9--dc22   2008011643

Visit the Taylor & Francis Web site at
http://www.taylorandfrancis.com

and the CRC Press Web site at
http://www.crcpress.com

Contents

Preface vii
Acknowledgments ix
Introduction xi

Chapter 1   Imperfections of the Crystal Structure 1

1.1   Dislocations 1
1.2   Multiplication of Dislocations 11
1.3   Vacancies 15
1.4   Grain and Subgrain Boundaries 19
1.5   Twins 25

Chapter 2   Formation of a Dislocation Substructure by Plastic Deformation 31

2.1   Formation of a Dislocation Substructure at Room Temperature in
       Metals of High Stacking Fault Energy 33
2.2   Formation of a Dislocation Substructure at Higher Temperature in
       Metals of High Stacking Fault Energy 39
2.3   Formation of a Dislocation Substructure at Room Temperature in
       Metals of Medium-Low Stacking Fault Energy 43
2.4   Formation of a Dislocation Substructure at Room Temperature in
       Metals of Very Low Stacking Fault Energy 49
2.5   Formation of a Dislocation Substructure at Higher Temperature in
       Metals of Low Stacking Fault Energy 55

Chapter 3   Changes in the Deformation Structure Caused by Heating 59

Chapter 4   Growth of the Crystals and Rapid Solidification 69

4.1   Growth of the Crystals 69
4.2   Rapid Solidification Process 73

Chapter 5   Solid-State Phase Transformations 79

5.1   Continuous Precipitation in Age-Hardening Alloys 80
5.2   Interaction of Dislocations with Second-Phase Precipitates 93
5.3   Discontinuous (Cellular) Precipitation 97
5.4   Eutectoid Transformation 103
5.5   Martensitic Transformations 109
5.6   The Bainite Transformation 121

Chapter 6   Case Studies: Application of TEM in the Solving of
                Problems in Engineering Practice 127

6.1   Deformation Behavior of Nickel-Silver Alloy in the
       Temperature Range from 100°C to 900°C 127
6.2   Distribution of Strengthening Phases in Precipitation-Hardening Alloys ... 133
6.3   Specific Features of the Structure Developed during Deformation in
       Superplastic State 141
6.4   The Influence of Modification and Heat Treatment on the
       Microstructure of Aluminum-Silicon Alloy 147
6.5   Sigma-Phase Formation in a Duplex Stainless
       Chromium-Manganese-Nitrogen Steel 151
6.6   Corrosion Resistance of Particular Structure Components of
       Austenitic Stainless Steels 155
6.7   Characterization of Ferrite in Welds of Austenitic Steels 163

Recommended Literature for Further Reading 169

Index 171

Preface

The contemporary teaching in the field of materials science, which encompasses physical metallurgy as an integral part, is based on the structure-properties-processing-performance relationship. With this idea in mind, as well as based on the academic teaching experience of the authors, this book was written to complement and visualize a number of topics included in basic academic courses in materials science, physical metallurgy, and phase transformations and in a number of excellent books.

This book is a teaching aid, designed as an atlas that comprises a collection of original transmission electron micrographs contributed by the authors. A JEM 7-A transmission electron microscope (TEM) of JEOL Company-Japan was used to characterize the microstructure. The micrographs were carefully selected and integrated with the purpose of demonstrating typical crystal lattice defects, elements of the microstructures of metals and alloys, and the basic processes occurring in the crystal structure during plastic deformation, polygonization, recrystallization, heat treatment, and rapid solidification. Considerable attention was given to the nanostructural features that can be visualized by the TEM and that represent the basis of solid-state reactions and transformations. Thus, the reader will be able, in a step-by-step fashion, to interpret TEM images both correctly and easily, as well as better understand the processes occurring in metallic structures at the nanolevel.

This book is organized into six chapters. Each chapter deals with a particular problem in the field of physical metallurgy and starts with a short description of the basic concepts and terms in order to enable the reader to achieve a better understanding of the essential issues related to that problem. We attempted to emphasize the most characteristic elements of the microstructure for each of the particular phenomena or class of materials rather than to illustrate a microstructure of a specific metal or alloy grade. Nevertheless, many of the selected microphotographs were taken from the commercial alloys samples. In addition, Chapter 6 presents an organized set of micrographs, demonstrating the scope of TEM as an experimental tool in the wider context of microstructural investigations applicable in practice to assist the solution of particular technological problems and to improve the service behavior of metallic materials.

It is our sincere hope that this book will provide a useful reading to undergraduate, graduate, and PhD-level students in materials science, metallurgy, and mechanical engineering departments and that they will benefit from its use as a teaching aid. We also hope that it will contribute to the broadening of the knowledge of anybody else that is interested in this topic or is professionally working in the field of physical metallurgy and materials science.

The authors

Acknowledgments

We would like gratefully to acknowledge the following individuals, who gave us invaluable help during the preparation of this book:

Dr. Michael Witcomb from the Electron Microscope Unit, University of the Witwatersrand, South Africa, for his expert advice and editing of the text, as well as for presenting our book to the publisher.

Dr. Lilian Ivanchev, formerly from Sofia University of Chemical Technology and Metallurgy, now employed in CSIR, Pretoria, for being supportive throughout the overall work on this book and in the less wonderful moments.

Staff of the TEM Laboratory at the Institute of Metal Science, Bulgarian Academy of Sciences, for their technical expertise and efficient help.

All the colleagues and students in the field of materials science and metallurgy who have been our collaborators over the years and who gave to us their encouragement and suggestions in writing this book; they are too numerous to list. Special thanks are due, however, to Dr. Margarita Gancheva (University of Chemical Technology and Metallurgy, Sofia) and Dr. Tzanka Kamenova (Institute of Metal Science, Bulgarian Academy of Sciences), who evaluated the manuscript in Bulgarian and offered fruitful suggestions and encouragement.

The editorial team at Taylor & Francis for their kindness and all the hard work they put into the preparation and editing of the manuscript and the production of the book.

Introduction

The source of illumination in a transmission electron microscope (TEM) is a beam of electrons emitted by the TEM source and accelerated in a high-voltage field (up to 1000 kV) in vacuum. The energy of the electrons is high enough for a part of them to pass through thin regions of the object placed in their path. These *transmitted electrons* then can be imaged on a fluorescent screen or CCD camera located below the object. The great advantage of TEM lies in the possibility of observing very small features (down to the order of several nanometers) at very high magnifications (up to hundreds of thousands or even millions of times). This is due to the high resolving power of the TEM, which is thousands of times greater than the resolving power of a light microscope (typically several tenths of a micrometer). The resolution of the TEM is drastically improved by the smaller wavelength of electrons (about 0.004 nm at 100 kV accelerating voltage), compared to the wavelength of light (about 1000 nm). However, the ultimate resolution of the microscope is limited by the lens aberrations. With current technology, a resolution down to sub–0.1 nm is possible.

Two types of specimens are used for the TEM study of structure of metals and alloys: replicas and thin foils. The *surface replica* reprints the surface relief of an etched metallographic specimen, of a working or fractured surface. The *extraction replica* pulls out precipitates from an etched sample so that, for example, the composition of the inclusions can be determined without any influence from elements within the surrounding matrix material. The *thin foil* is a metal slice prepared from bulk material so thin (typically several hundreds of nanometers) that it is transparent to the accelerated electrons in the TEM.

The contrast in a TEM image results from the interaction of the beam of accelerated electrons with the electrons of the object (specimen). When penetrating the specimen, the accelerated electrons undergo *electron absorption* by an amount depending on the target thickness and atomic number. When penetrating a periodic structure (crystalline material), some of the electrons undergo *electron diffraction*. These electrons are referred to as *diffracted electrons*; the electrons, which pass through the thin crystal without being diffracted, are termed *transmitted electrons*. The different ratio of transmitted to diffracted electrons passing through areas with different crystallographic characteristics causes *diffraction contrast*. This type of contrast is especially useful in studying the imperfections in metals and alloys: dislocations, complexes of point defects, stacking faults, grain and twin boundaries, precipitate interfaces, and local stress fields.

Electron diffraction offers an additional means for crystallographic analysis of the crystal structure. The electrons diffracted from different planes of the crystal lattice form *diffraction patterns,* which can also be recorded in the TEM. Each *diffraction spot* in the pattern corresponds to a certain reflecting crystal plane, and the analysis of the pattern size and geometry provides information regarding the structure type and orientation of the crystal.

The main linear imperfections in the structure of real crystals are the *dislocations*. The dislocation line separates a region of the crystal in which crystallographic slip has occurred from the adjacent unslipped region. The dislocation is imaged in the TEM through the diffraction contrast arising from the change of atomic arrangement at the border of the two regions. The image of the dislocation in bright field imaging is a dark line, which can be visible or invisible depending on the diffraction extinguishing conditions; these are determined by the angle between the Burgers vector of the dislocation and the incident beam direction.

Vacancies are point defects—crystal lattice sites not occupied by atoms. A single vacancy cannot be observed by TEM, because its size is smaller than the TEM resolving power. Vacancies clustered together in planar complexes—*vacancy loops (discs)*—can be observed through the presence of a dislocation line in the form of a loop bordering the complex. When the vacancies collect together in a three-dimensional form, they are termed *voids*.

High-angle boundaries and *stacking faults* are two-dimensional (planar) imperfections in crystal structures at which the periodic arrangement of crystal planes changes. The planar defects are visualized in the TEM due to the difference in phase of the electron beams diffracted from above and below the plane of the imperfection. The interference of the two diffracted beams produces characteristic *interference fringes*—alternating black and white lines (fringes)—parallel to the intersection of the defect with the upper and lower surfaces of the sample foil.

The elastic distortion or shift of atomic layers caused by the local *strain fields* present in the crystal structure changes the geometry of the transmitted electrons and causes *deformation contrast* in the TEM in the form of diffuse dark shadows.

1 Imperfections of the Crystal Structure

1.1 DISLOCATIONS

Dislocations are line defects of the crystal lattice, which border the regions in the crystal interior in which slip has occurred. The measure for the quantity of dislocations in the crystal is given as the *dislocation density*—the total length of all of the dislocation lines per unit volume.

The main characteristic of each dislocation is the *Burgers vector*, which describes the magnitude and direction of slip movement associated with the dislocation. The classification of dislocations is based on the mutual orientation of the dislocation line and the Burgers vector of the dislocation. The direction of an *edge dislocation* is perpendicular to the Burgers vector. In contrast, the direction of a *screw dislocation* line is parallel to the Burgers vector. Most of the dislocations are *mixed dislocations*. They are a combination of screw and edge segments, as well as a large mixed component; that is, the direction of the Burgers vector of such a dislocation changes along the length of the dislocation. Mixed dislocations usually form *dislocation loops*.

When the magnitude of the Burgers vector equals a whole lattice vector, it is referred to as a *unit* or *perfect dislocation*. Such dislocations, when passing through the crystal, do not change the arrangement of atoms in the lattice because a complete lattice translation occurs. Dislocations with a Burgers vector not equal to a whole lattice vector are referred to as *imperfect* or, more often, as *partial dislocations*.

Any perfect (unit) dislocation may have an edge, screw, or mixed nature; each mixed dislocation can be either unit or a partial. Each type of close-packed crystal structure has its own set of unit and partial dislocations.

All dislocation reactions are described by the magnitude and direction of the Burgers vectors. The *dislocation reactions* in real crystals are as follows:

- *Combining* of two or more dislocations
- *Splitting* of unit dislocations into partial dislocations
- *Interaction* of partial dislocations resulting in dislocations of a new type
- *Annihilation* of dislocations with opposite Burgers vectors

A basic physical characteristic of metals and alloys that governs the mode of dislocations movement and the type of dislocation reactions is the *stacking fault energy (SFE)*—the energy necessary to produce a unit area of *stacking fault (SF)* in a perfect crystal. The SF is a planar lattice disorder with a stacking sequence of layers different from the perfect close-packed structure: for example, a layer of ...CACA... stacking sequence characteristic for hexagonal close-packed (hpc) crystal lattice inserted into

1

the face centered cubic (fcc) lattice with …ABCABC… stacking sequence results in …ABCACABC… arrangement, which contains an hpc layer of stacking fault.

The area of the SF is bordered by two *partial dislocations*: a leading one, which disturbs the stacking sequence, and a closing one, which restores the regular atomic arrangement of the matrix. The complex of the two partial dislocations connected by SF ribbon is called *split dislocation*, the equilibrium *width of the split dislocation* (distance between the two partials) determined by the SFE of the crystal. The value of SFE determines also the type of *dislocation substructure* produced in materials by plastic deformation.

The stacking faults are the simplest type of planar defects. When they are inclined to the foil surface, they are visualized in the TEM as ribbons containing parallel white and black fringes resulting from the diffraction contrast arising from the planar imperfection in the periodic arrangement of the atoms.

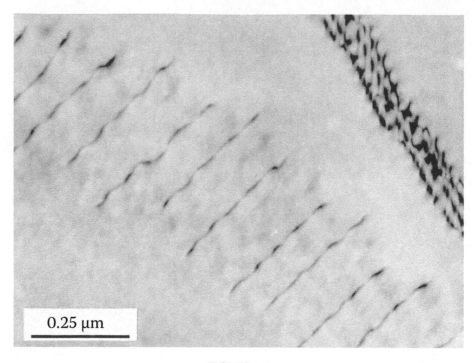

0.25 μm

FIGURE 1.1

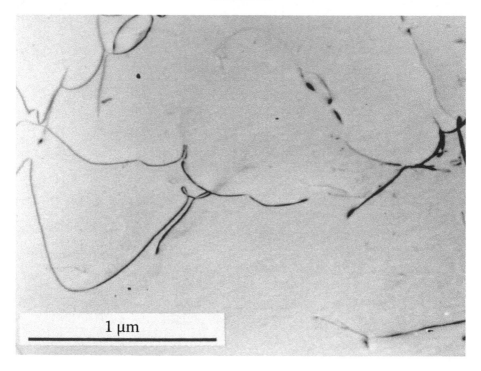

1 μm

FIGURE 1.2

FIGURE 1.1   Real crystals contain a significant number of dislocations. An annealed crystal will have a dislocation density of approximately 10^7 cm of dislocation length per cubic centimeter of material (or 10^7 cm^{-2}). The smaller the number of dislocations per unit volume, the longer the single dislocation line (a sugar-cube-sized piece of any engineering alloy contains about 10^5 kilometers of dislocation line). The length of a single dislocation line visible in the TEM is much shorter: for example, less than 1 micrometer in the shown micrograph. This is because the dislocations are inclined to and thus cut by the two foil surfaces so that we see only short segments of them.

Quenched stainless austenitic steel. The oscillatory contrast along the shown dislocation segments is caused by the interference between the diffracted and transmitted electron beams near to the surfaces of the thin foil.

FIGURE 1.2   The dislocations in crystals of high SFE are always unit (perfect) dislocations. They can move by slip, not only in the plane of their Burgers vector but also in the intersecting planes, to produce a *three-dimensional* dislocation distribution. This type of distribution is characteristic of many pure metals—nickel, chromium, molybdenum (SFE about 300 mJ·m^{-2}), aluminum (SFE = 250 mJ·m^{-2}), magnesium (SFE = 200 mJ·m^{-2}), titanium (SFE = 150 mJ·m^{-2}), α-iron (SFE = 140 mJ·m^{-2}), and zinc (SFE = 100 mJ·m^{-2}).

Quenched aluminum.

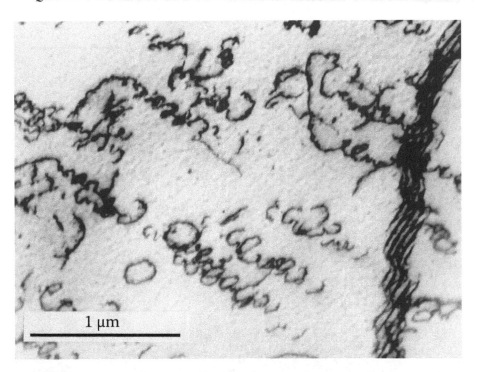

1 μm

FIGURE 1.3   When a large number of vacancies are present in metals and alloys of high SFE, they produce *helical dislocations*. It is energetically advantageous for vacancies to condense around the axis of screw dislocations. Climbing dislocations with vacancies condensed around them obtain the shape of regular helices with an axis along the Burgers vector of the original screw dislocation. This process can often be observed in aluminum alloys quenched from a high temperature where it is an indicator of a high vacancy supersaturation in the alloy.

Quenched Al-4% Cu.

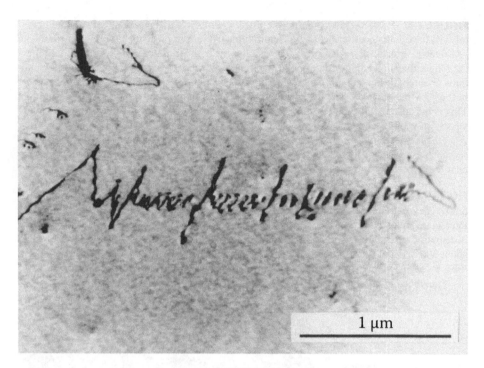

FIGURE 1.4

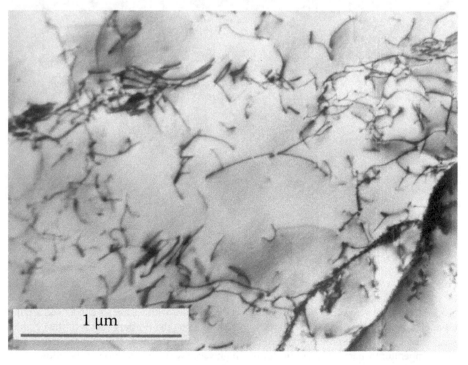

FIGURE 1.5

FIGURE 1.4 The shown image is not an ancient Arabian script, but a helical dislocation.
Quenched Al-4%Cu.

FIGURE 1.5 A lower SFE makes it difficult for dislocations to slip in planes different from the plane containing their Burgers vector. Nevertheless, a three-dimensional distribution of dislocations is still observed in metals of SFE about 30 to 40 mJ·m^{-2}, for example, in copper or niobium. This is because the *cross-slip* to a plane other than the primary slip plane is still possible as a result of the small separation of the partial dislocations.
Quenched copper.

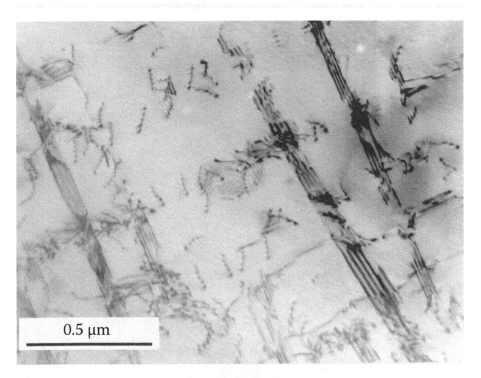

0.5 μm

FIGURE 1.6 The distribution of dislocations in materials having an SFE less than 20 mJ·m^{-2} is strictly *two-dimensional* or *planar*. A representative pure metal of this type is silver with an SFE of 15 mJ·m^{-2}. The planar distribution of dislocations is more frequently observed in alloys because alloying usually lowers the SFE. Examples are the stainless steel Fe-18Cr-9Ni (SFE = 16–40 mJ·m^{-2}), α-brass CuZn30 (SFE = 10 mJ·m^{-2}), and aluminum bronzes with 4% to 7% Al (SFE = 3–5 mJ·m^{-2}).
Quenched stainless austenitic steel Fe-18Cr-9Ni.

FIGURE 1.7

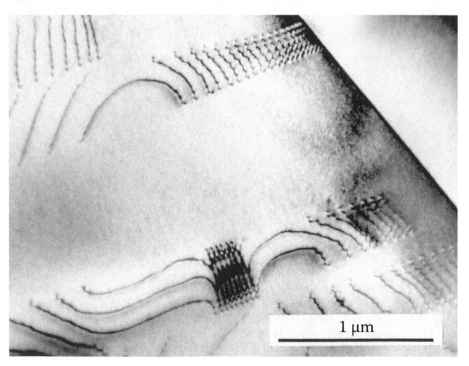

FIGURE 1.8

FIGURE 1.7 When the value of the SFE is very small, it is energetically more advantageous for dislocations to move by splitting into partials separated by SF. The lower the SFE value, the wider the SF layer: in the case of an SFE of about 1 mJ·m⁻², the length of the SF layer can reach several micrometers.

Quenched austenitic nitrogen steel Fe-18Cr-14Mn-0,6N. The change of brightness or the loss of SF image in some areas in the micrograph results from the presence of several overlapping SFs in the foil thickness, which causes a change in the transmitted electrons' intensity.

FIGURE 1.8 By changing the imaging conditions used in Figure 1.7, the parallel black and white fringes typical of a SF image can be made invisible.

Quenched austenitic nitrogen steel Fe-18Cr-14Mn-0,6N. As in Figure 1.7, the partial dislocations in this micrograph are arranged in arrays in the parallel slip planes and actually are connected by SF bands, but the imaging conditions make these bands invisible.

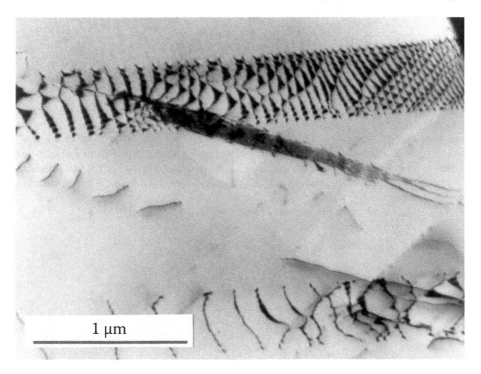

1 µm

FIGURE 1.9 The intersection of two widely extended dislocations that move on different slip planes in a low-SFE metal produces a *triple dislocation node* containing an SF. Under equilibrium conditions, the area of the node, bordered by partial dislocations, serves as a measure of the SFE value.

Quenched austenitic nitrogen steel Fe-18Cr-14Mn-0,6N. The specific network of alternating expanded and shrunk triple nodes in the shown micrograph is a result of the crossing of two arrays of split dislocations moving on two intersecting planes.

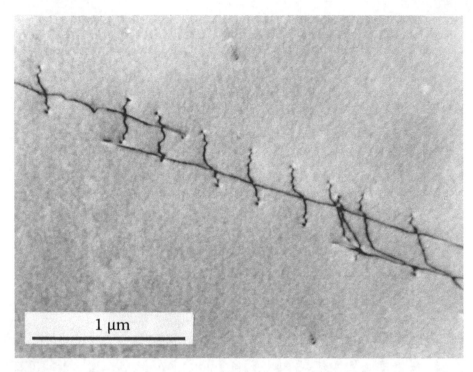

FIGURE 1.10   When a pair of extended dislocations meet at the intersection of two slip planes, the two leading partial dislocations interact and produce a new partial dislocation, which is referred to as a *stair-rod dislocation*. The stair-rod dislocation (named after the rod that holds a stair carpet in place) is practically immobile because of the complex structure of the three dislocations connected by a common wedge-shaped SF. This is the so-called *Lomer-Cottrell barrier* or *sessile dislocation*. It is recognized as one of the most difficult obstacles for dislocations to surpass and thus plays an important role in the work hardening of the fcc metals.

Quenched austenitic nitrogen steel Fe-18Cr-14Mn-0,6N.

1.2   MULTIPLICATION OF DISLOCATIONS

The number of dislocations in a crystal is changed by mechanical processing. The density of dislocations in a well-annealed crystal varies from 10^6 to 10^8 cm^{-2}. After 30 to 40% cold plastic deformation, the dislocation density increases to 10^{11}–10^{12} cm^{-2}. One of the main sources for the multiplication of dislocations during plastic deformation is a *Frank-Read source*. A Frank-Read source is a region of the crystal with a high density of defects, which is capable of generating dislocations when the shear stress reaches a certain critical level. The process is repeated when a new dislocation breaks out and starts moving away from the source. The source continues to emit dislocations of the same type and can generate an unlimited number of dislocations if the applied stress remains in excess of the critical value.

The dislocations moving through the crystal have to overcome numerous obstacles—point defects and defect clusters, foreign atoms or phases, inhomogeneities and stress fields associated with the solid solution, other dislocations or dislocation complexes, and grain and twin boundaries. If the obstacle is difficult to overcome—for example, a grain boundary, an inclusion interface, or a field of stress concentration—the dislocations stop and form *dislocation pileups*. The stress on the leading dislocation produced by the pileup is proportional to the number of dislocations in the pileup. In many cases, it is sufficient to activate a Frank-Read source.

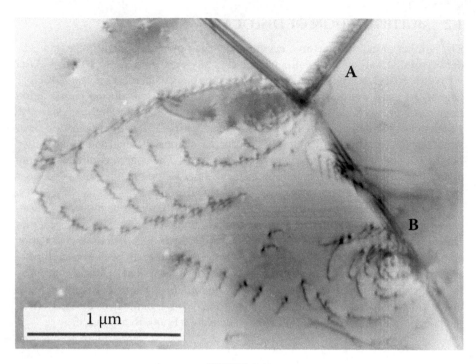

FIGURE 1.11

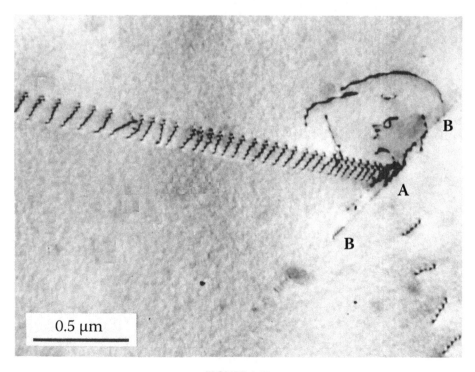

FIGURE 1.12

FIGURE 1.11　Frequently, grain boundaries contain Frank-Read sources.

Quenched austenitic nitrogen steel Fe-18Cr-14Mn-0,6N. Two grain boundary sources are visible in this micrograph: at the triple point A and at point B. Both sources emit dislocations into the grain on the left.

FIGURE 1.12　Dislocation-type barriers contain Frank-Read sources also. Such a barrier is shown at point A in the micrograph. The movement of the large dislocation pileup on the left has been stopped by a plane of high dislocation density intersecting the foil surface at B-B. The stress on the leading dislocation of the large pileup (it contains more than 40 dislocations) has activated a Frank-Read source. The emitted new dislocations glide in two directions—to the top and to the bottom of the micrograph.

Quenched austenitic nitrogen steel Fe-18Cr-14Mn-0,6N.

1.3 VACANCIES

The sites in a crystal lattice that are not occupied by atoms are called *vacancies*. The energy necessary for the formation of a vacancy is very small—about 1 eV—which explains the large concentration of these point defects under thermodynamic equilibrium conditions in metals and alloys. *Sources* for the formation of vacancies are the free surfaces of the crystals and the internal defects (dislocations, grain and subgrain boundaries, phase interfaces).

Vacancies in excess of their equilibrium concentration are generated most often during quenching from high temperatures, during plastic deformation, during ion bombardment, during bombardment by high-energy nuclear particles, or, in some intermetallic compounds, as a result of stoichiometric deviations. They are also produced by the oxidation of some metals, such as Mg, Ni, Cu, Zn, and Cd.

The equilibrium concentration of vacancies increases exponentially with temperature and is very high at the solution treatment temperature. When a solution-treated alloy is rapidly cooled (quenched) from a high temperature, the number of vacancies in the structure remains much higher than the lower-temperature equilibrium concentration. The quenched-in vacancies determine the level of the *vacancy supersaturation*. When the rate of cooling from the solution treatment temperature is lower, vacancies have sufficient time to reach *vacancy sinks* (sites in the lattice where they are annihilated), thus reducing the vacancy supersaturation.

Vacancies play a significant part in nonconservative dislocation movement (climb) and in the processes involving a diffusion transport of atoms—for example, polygonization and recovery. They thus play an important role during the processing of metals and alloys by plastic deformation at elevated temperatures; during solid solution heat treatment, annealing and aging; during creep; as well as in the processes of irradiation damage caused by high-energy particles.

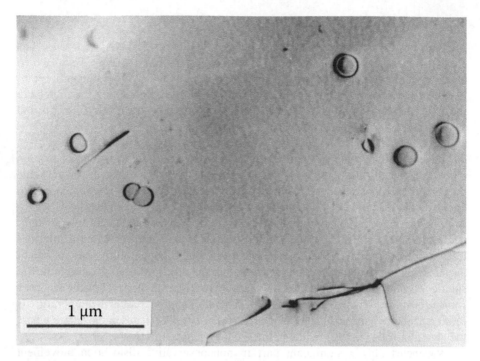

1 µm

FIGURE 1.13

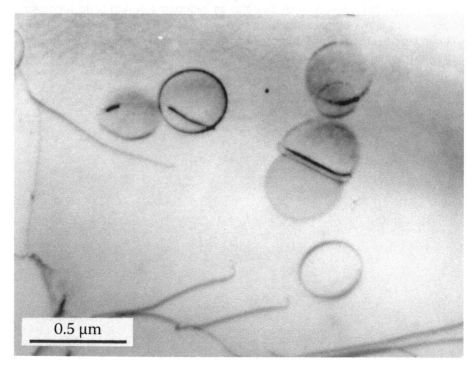

0.5 µm

FIGURE 1.14

FIGURE 1.13 Zinc is characterized by its very pronounced ability to form vacancies through surface absorption processes. The vacancies combine into disc-shaped complexes in the basal plane of the hexagonal close packed (hcp) crystal lattice. The separate vacancies are invisible in the TEM because of their small size, but the presence of the *dislocation loops* that border the complexes make the vacancy discs visible.

Quenched zinc.

FIGURE 1.14 The number of vacancies in quenched zinc crystals is high. The bombardment of the thin metal foil by accelerated electrons during TEM observation introduces an additional number of point defects in the surface region and leads to the significant enlargement of the disc-shaped vacancy complexes. The process is similar to that of irradiation-induced damage occurring in industrial nuclear materials.

Quenched zinc after prolonged observation in the TEM. Compare this micrograph to Figure 1.13.

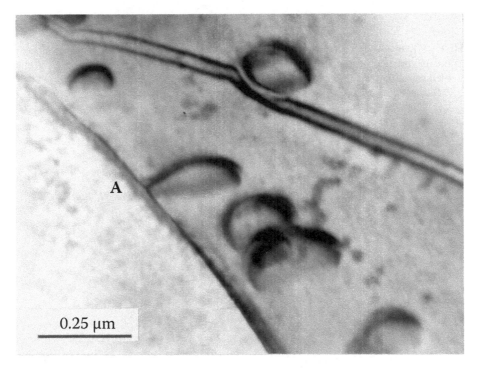

FIGURE 1.15 Besides acting as sources for vacancy formation, the imperfections in the crystal structure (mainly grain boundaries and dislocations) serve as sinks for their annihilation.

Quenched zinc after a prolonged holding in the evacuated column of the TEM. The grain boundary serves as a sink for the vacancies of the discs lying in close proximity to it. One can even see the "flowing out" of the vacancies to the boundary at point A.

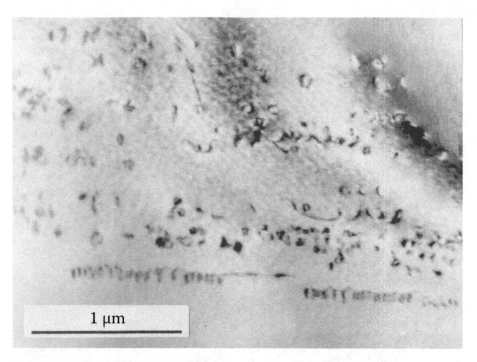

FIGURE 1.16 A high density of vacancy complexes (*vacancy loops*) is a typical component of the microstructure associated with quenched aluminum and aluminum alloys. This is due to their capacity to achieve a high vacancy supersaturation.

Quenched Al-4%Cu.

1.4 GRAIN AND SUBGRAIN BOUNDARIES

The *boundaries* of grains belong to the group of two-dimensional lattice imperfections. They are interfaces that separate regions in the interior of the material at which the crystal lattice changes orientation. The type and nature of the boundaries depend on the *misorientation angle* of the two adjoining grains and on the orientation of the interface boundary plane to them.

A *low-angle boundary (subboundary)* is a wall of tangled dislocations. The misorientation angle across the boundary depends on the number of dislocations that build up the dislocation wall. In a low-angle boundary, the angle does not exceed several degrees.

The misorientation angle of a *high-angle boundary* ranges from several degrees and can reach several tens of degrees. When the angle exceeds 10 to 15 degrees, the boundary is known as a *random high-angle boundary*. This boundary of several interatomic distances in width is in a high state of disorder compared to the matrix crystal structure—the atoms are out of their normal positions, the interatomic bonds are distorted, and consequently the boundary is associated with a higher energy.

There are many models (e.g., dislocation and disclination models, model of the coinciding knots) describing the structure of high-angle boundaries as a function of their misorientation angle, but no common theory has been accepted up to now.

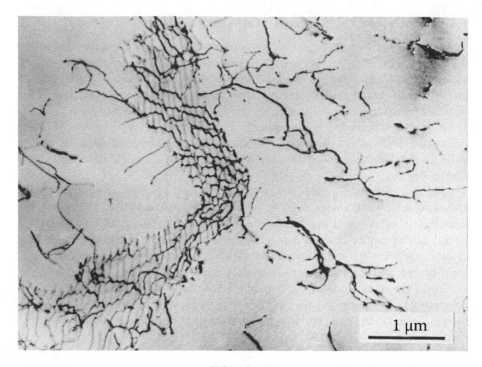

FIGURE 1.17

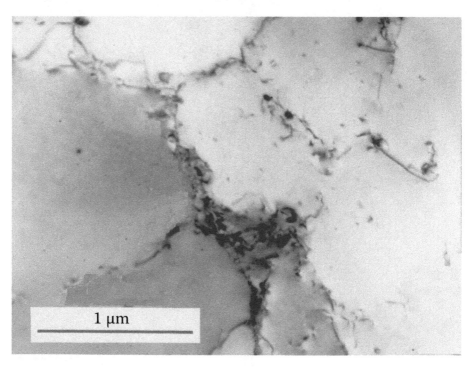

FIGURE 1.18

FIGURE 1.17 The dislocation complexes, called *tangles*, are produced by the interaction of moving dislocations. When the number of tangled dislocations becomes higher, they build a *dislocation wall*.

Annealed aluminum.

FIGURE 1.18 When reaching the dislocation wall, the moving free dislocations interact with the wall dislocations. This interaction results in more regular and stable dislocation complexes—*subboundaries* that separate *subgrains* regions of the crystal with relatively low dislocation density. Each dislocation participating in subboundary formation contributes to the misorientation of the adjacent subgrains by a magnitude depending on its Burgers vector. The amount of misorientation can be estimated in TEM images by the white-black contrast across the boundary: the stronger the contrast, the larger the misorientation angle.

Subgrains in aluminum.

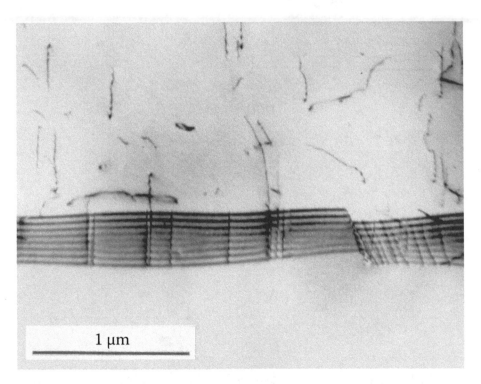

1 μm

FIGURE 1.19 High-angle boundaries are planar defects in the crystal structure. When the plane of the boundary is inclined toward the foil surface, the boundary is seen in the TEM as a ribbon with parallel sides, the latter being the intersections of the grain boundary with the two foil surfaces. Depending on the TEM imaging conditions, parallel black-white fringes typical for the TEM images of all planar defects can be visible or invisible inside the ribbon.

Quenched austenitic nitrogen steel Fe-18Cr-14Mn-0,6N. Several parallel dislocations appear to be moving inside the boundary in the micrograph, but they are actually moving in a plane either above or below the boundary.

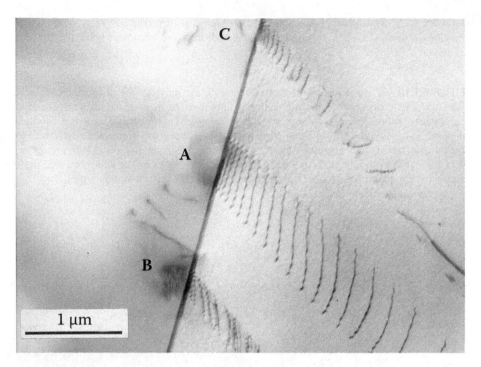

FIGURE 1.20

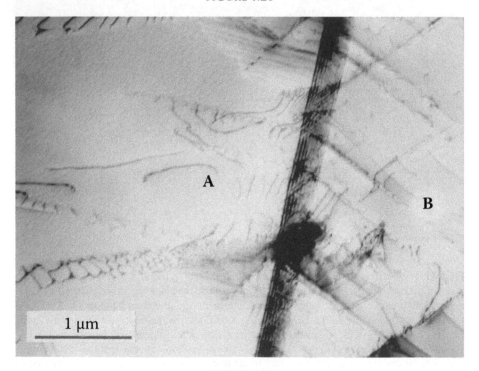

FIGURE 1.21

FIGURE 1.20   The high-angle boundaries are practically impenetrable barriers to moving dislocations.

Quenched austenitic nitrogen steel Fe-18Cr-14Mn-0,6N. The distance between the dislocations in the shown pileups when approaching the boundary diminishes. The stress on the leading dislocations, which can be estimated by the dark shadows at A, B, and C, is very high. When the stress reaches a critical level, the dislocation sources in the boundary will be activated to produce dislocations in the grain on the left.

FIGURE 1.21   The initiation of new dislocations in the next grain is the base mechanism for dislocation transfer across the boundaries during the plastic deformation.

Austenitic nitrogen steel Fe-18Cr-14Mn-0,6N after 2% cold deformation.

1.5 TWINS

A *twin* is a part of the crystal that has a crystal lattice identical to that of the base crystal *(matrix)*, but with a crystallographic orientation that is a mirror image of the matrix orientation. Twins are always bordered by a pair of parallel *coherent boundaries* at which is realized the symmetrical tilt between the two twin-related crystals. Because of their extremely low energy due to the perfect fitting of boundary atoms into the lattices of both grains, the coherent boundaries are regarded as special high-angle grain boundaries. The other boundaries of the twin, which have a random orientation to the twinning plane and no geometric relationship to the matrix, are *incoherent boundaries*. The energy of these boundaries is much higher because the boundary atoms do not fit exactly into each grain lattice.

An important feature of twins is the fact that they never cross grain boundaries; they can terminate at grain or twin boundaries or inside the grain interior.

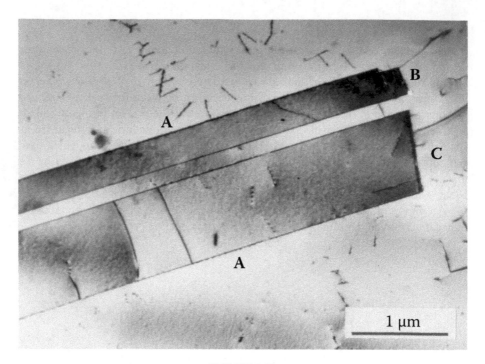

FIGURE 1.22

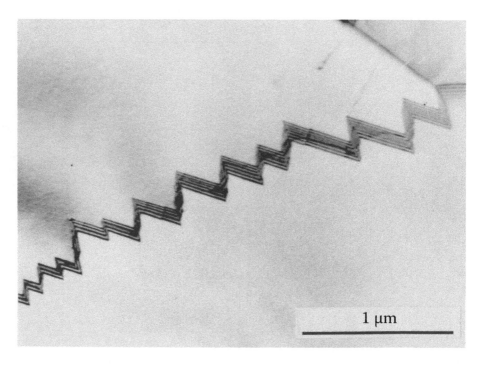

FIGURE 1.23

FIGURE 1.22   The orientation of the different twins in the same grain is usually identical.
   Quenched austenitic nitrogen steel Fe-18Cr-14Mn-0,6N. The coherent boundaries (A) of
the two twins shown in the micrograph are parallel. The incoherent boundaries (B and C) are
randomly oriented to the matrix.

FIGURE 1.23   The grain boundary shown in the micrograph in Figure 1.23 contains coherent (parallel) and incoherent (nonparallel) steps.
   Quenched austenitic nitrogen steel Fe-18Cr-14Mn-0,6N.

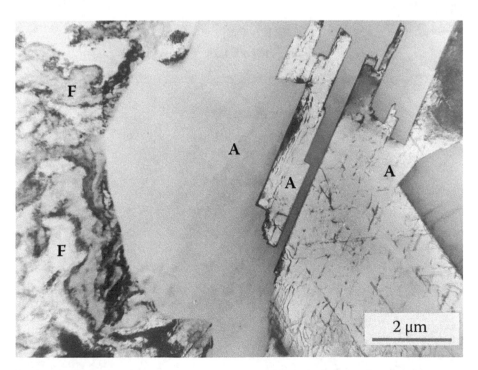

FIGURE 1.24   Twinning is a typical process occurring during recrystallization of low-SFE
metals and alloys. These twins are called *annealing twins.* The presence of annealing twins
in austenite helps to easily distinguish the austenite (A) from the ferrite (F) grains in a
ferrite-austenite (duplex) steel.
   Quenched ferrite-austenite steel.

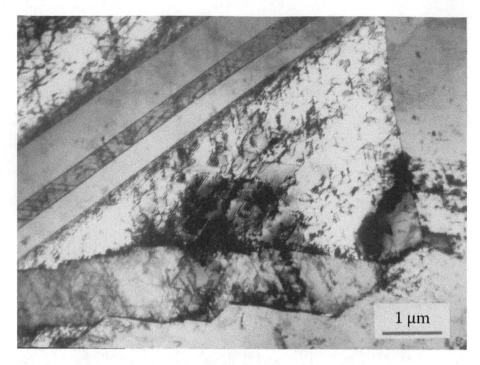

FIGURE 1.25

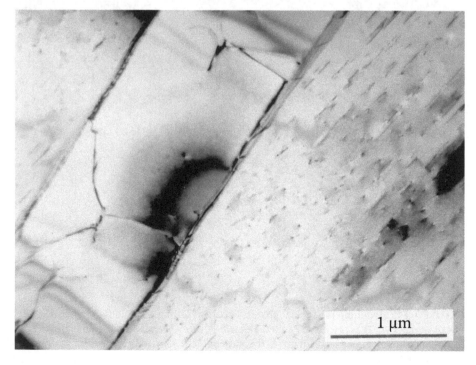

FIGURE 1.26

FIGURE 1.25   The frequency of twin formation helps to roughly assess the SFE value of metals and alloys. Twins can never be found in aluminum (high SFE), but they are a common feature in annealed copper (medium SFE) and are a compulsory element in recrystallized α_{cu}-grains of brass (very low SFE).
   Brass CuZn30.

FIGURE 1.26   *Mechanical twinning* is a principal deformation mechanism in metals having an hcp crystal lattice (Mg, Zn, Cd). This is due to the restricted slip, which can occur only in the basal plane. The twinning process in some of these metals is accompanied by energy release in the form of sound—for example, the well-known "metal yield" accompanying the bending of a tin rod due to the rapid formation of numerous twins.
   Zinc after 2% deformation.

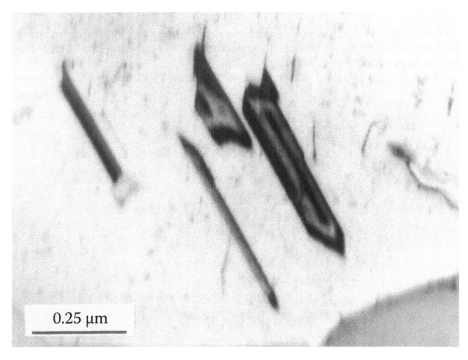

0.25 μm

FIGURE 1.27   Twinning is a comparatively rare process in body-centered cubic (bcc) crystals. Only TEM observations can reveal the extremely narrow *microtwins* formed in some low-alloyed steels.
   Ferritic steel containing 10%Mn, 2%Si, 0.1%V, and 0.1%N.

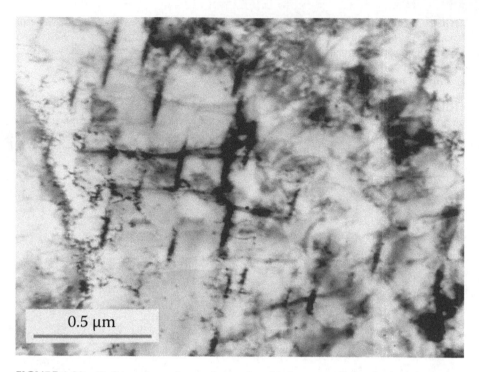

0.5 µm

FIGURE 1.28 *Deformation twins* can be produced in bcc crystals by plastic deformation, especially through dynamic loading. These mechanical twins are usually very thin and are referred to as microtwins.

Ferritic steel containing 10%Mn, 2%Si, 0.1%V, and 0.1%N after high-velocity plastic deformation.

2 Formation of a Dislocation Substructure by Plastic Deformation

The plastic deformation of metals and alloys occurs by the elementary processes of nucleation, movement, interaction, and annihilation of crystal lattice imperfections (vacancies, dislocations, grains, and twin boundaries). The dislocations start moving when the applied stress exceeds the yield strength of the metal. Dislocations of opposite sign may annihilate upon meeting, whereas the rest of the dislocations interact and combine into *dislocation complexes* or *tangles*. The increasing number of moving and interacting dislocations results in barely movable complexes, which form a *dislocation substructure*.

The character of dislocation substructure depends on several factors: the deformation conditions (temperature, strain rate, amount of deformation), the crystal structure, and the SFE of the metal or alloy. The strain rate and temperature are complementary variables—the effect of a strain rate increase is equivalent to a temperature decrease. Besides the diffusion-assisted processes, the temperature influences the type of dislocation reactions because of the strong temperature dependence of the SFE value. In the case of metals and alloys, which don't undergo temperature-induced phase transformations, the value of the SFE increases with the increasing temperature.

The dislocation substructure is of special importance, not only for the metal behavior under deformation, but also for the properties of the deformed metal.

2. Formation of a Dislocation Substructure by Plastic Deformation

2.1 FORMATION OF A DISLOCATION SUBSTRUCTURE AT ROOM TEMPERATURE IN METALS OF HIGH STACKING FAULT ENERGY

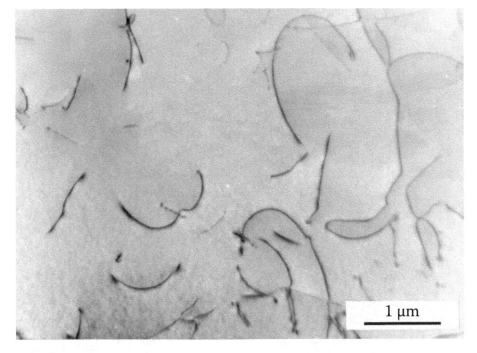

FIGURE 2.1 The density of dislocations in well-annealed metals and alloys is rather low. In metals and alloys of high SFE, such as aluminum, the unit dislocations can move on all slip planes of the crystal lattice conservatively (by slip) and nonconservatively (by cross-slip). The easy cross-slip of individual dislocations results in three-dimensional dislocation distribution. *Quenched aluminum.*

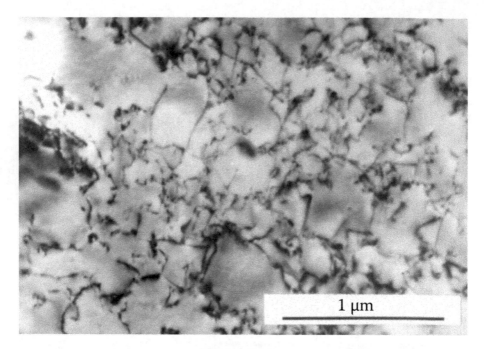

1 µm

FIGURE 2.2

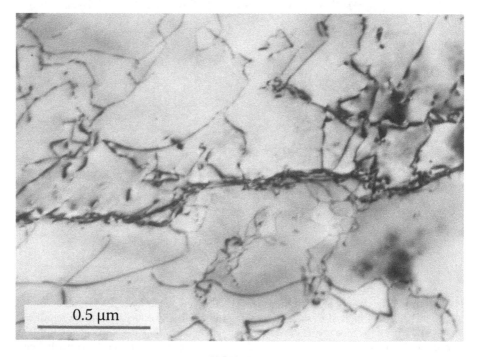

0.5 µm

FIGURE 2.3

FIGURE 2.2   As in the high-SFE aluminium, the dislocations in high-SFE ferrite have three-dimensional distribution in spite of the different crystal lattices, fcc for aluminium and bcc for ferrite.

Plain carbon steel, mould casting. The dislocation density in this case is higher than in the quenched aluminum, shown in Figure 2.1, because of the more complicated cooling conditions occurring during the casting.

FIGURE 2.3   When an external stress is applied to the crystal, the dislocations start moving, annihilating, multiplying and interacting, and trying to form configurations of maximum stability, that is, of minimum energy. For lower amounts of cold deformation, the number of moving dislocations is small and their interaction results in the simplest of complexes—dislocation tangles.

Aluminum after 2% cold deformation.

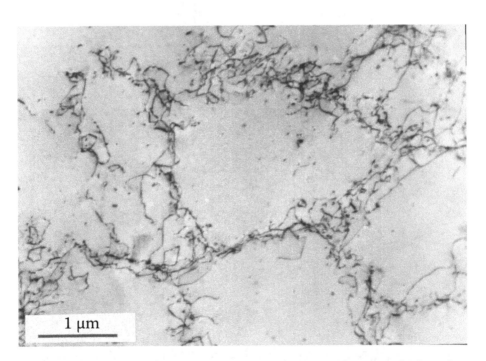

1 µm

FIGURE 2.4   The number of dislocations increases with increasing amount of deformation. The greater the number of dislocations that have participated in the formation of tangles, the more perfect and stable the tangles become. They gradually transform into *dislocation walls*—interconnected dislocation complexes—which divide the volume into *dislocation cells* with an interior almost free of dislocations. The term *cell structure* has been adopted in physical metallurgy for this type of cold-deformed structure in analogy to the term *cell structure* used for plants.

Aluminum after 5% cold deformation.

FIGURE 2.5

FIGURE 2.6

FIGURE 2.5   With increasing amount of deformation, the number of dislocations in cell walls increases and the size of the cells decreases, but the interior of the cells remains free of dislocations.
   Aluminum after 30% cold deformation.

FIGURE 2.6   After large amounts of deformation (over 50% reduction), the refinement of cells is accompanied by their elongation parallel to the deformation direction.
   Aluminum after 50% cold deformation.

2.2 FORMATION OF A DISLOCATION SUBSTRUCTURE AT HIGHER TEMPERATURE IN METALS OF HIGH STACKING FAULT ENERGY

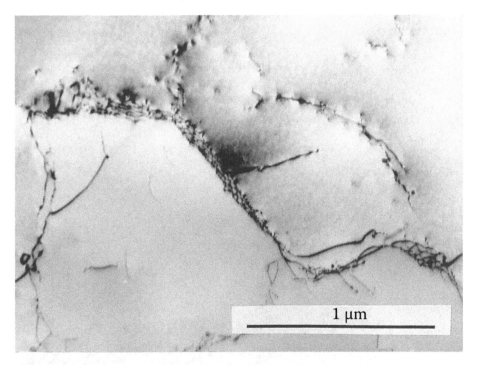

1 μm

FIGURE 2.7 When the plastic deformation is carried out at higher temperatures, the cross-slip of dislocations is assisted by diffusion processes. The annihilation of dislocations of opposite sign is thus much easier, and this reduces the dislocation density. Because the number of dislocations building the dislocation tangles is smaller, the tangles are narrower but more regular than at lower temperatures.

Aluminum after 2% deformation at 400°C.

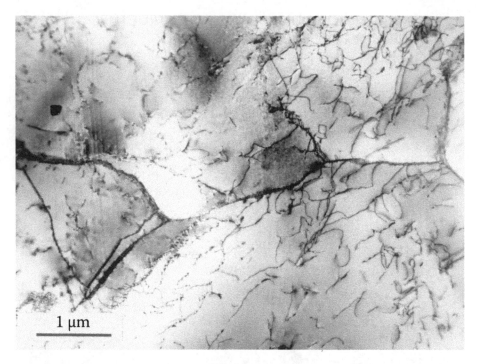

FIGURE 2.8

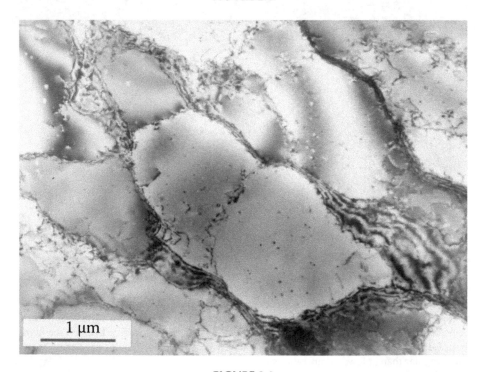

FIGURE 2.9

FIGURE 2.8 Each dislocation arriving at a cell wall contributes to the misorientation of the crystal lattice across the wall by a magnitude depending on the Burgers vector of that dislocation. At higher temperatures, the number of dislocations that succeed in passing through the free cell interior before meeting a wall is greater than at lower temperatures. When the magnitude of the misorientation between adjacent cells produced by these dislocations exceeds 1 to 2 degrees, the dislocation wall transforms into a *dislocation subboundary*.

Aluminum after 5% deformation at 400°C.

FIGURE 2.9 After a higher percentage deformation (more than 10%–15%) at a higher temperature, the three-dimensional dislocation tangles transform into two-dimensional dislocation complexes with a crystallographic misorientation of more than 10 degrees. The dislocation cell structure transforms into a *dislocation substructure* containing *subgrains* bordered by *high-angle subboundaries*. All metals and alloys of high SFE show a tendency to build a dislocation substructure when subjected to hot rolling or extrusion.

Aluminum after 15% deformation at 400°C.

2.3 FORMATION OF A DISLOCATION SUBSTRUCTURE AT ROOM TEMPERATURE IN METALS OF MEDIUM-LOW STACKING FAULT ENERGY

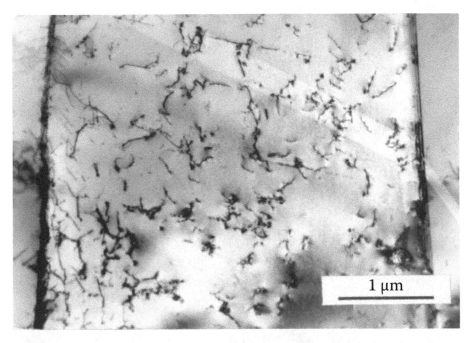

FIGURE 2.10   The splitting of dislocations in metals of low SFE inhibits cross-slip, but in metals of medium-low SFE (about 15–40 mJ/m^2), the width of SF separating the partials is relatively small. So the applied external stress can easily bring the partials together and cause them to behave as unit dislocation. This is the case in slightly deformed copper and in austenitic steel Fe-18Cr-9Ni, where the major portion of the dislocations move as unit dislocations and produce three-dimensional dislocation distribution.

Austenitic steel Fe-18Cr-9Ni after 2% cold rolling. There is only slight evidence for an arrangement of dislocations in planar rows.

FIGURE 2.11

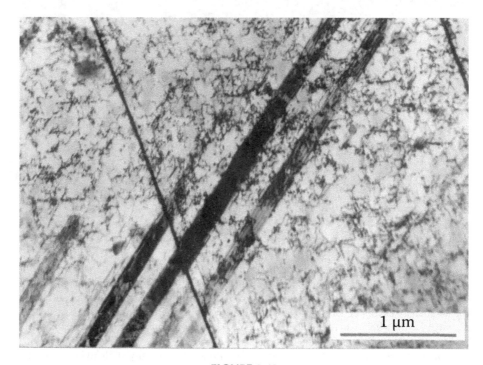

FIGURE 2.12

FIGURE 2.11   At moderate deformations, the dislocations in medium-low-SFE metals form tangles and imperfect dislocation walls but still preserve some tendency for planar arrangement.
Austenitic steel Fe-18Cr-9Ni after 10% cold rolling.

FIGURE 2.12   A frequent deformation mechanism in medium-low-SFE alloys is deformation twinning. In austenite, the twins form thin bands on the {111} crystal planes of the fcc lattice. The contrast produced by the deformation twins (the dark bands) is similar to the contrast arising from the bands of SF.
Austenitic steel Fe-18Cr-9Ni after 10% cold rolling.

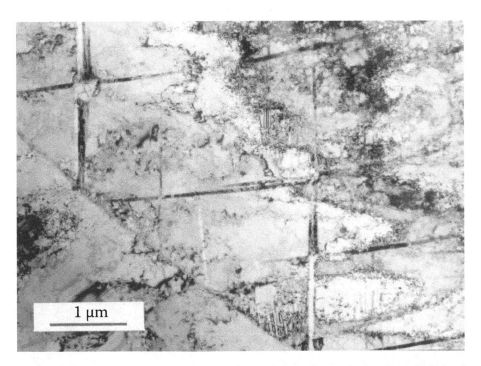

FIGURE 2.13   The substructure of a moderately deformed austenite consists of bands of deformation twins in a matrix containing refined dislocation cells.
Austenitic steel Fe-18Cr-9Ni after 20% cold rolling.

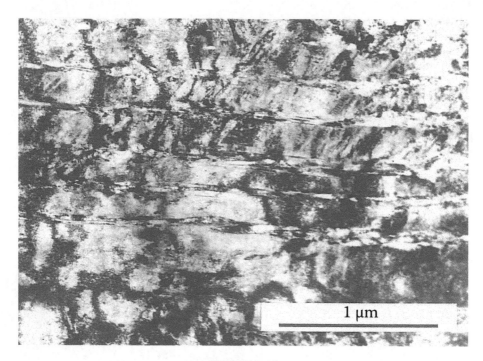

FIGURE 2.14

FIGURE 2.15

FIGURE 2.14 Deformation twinning develops only at moderate deformation amounts. The microstructure of severely deformed medium-low-SFE alloys contains extremely fine dislocation cells, in which individual dislocations are barely resolvable, and long straight microbands, which are due to the highly concentrated slip along the traces of the active slip planes. The microbands give rise to the well-known *slip lines*, which can easily be observed in a metallographic microscope.

Austenitic steel Fe-18Cr-9Ni after 50% cold rolling.

FIGURE 2.15 In some low-SFE steels—for example, the Hadfield steel—deformation twinning is the main mechanism of plastic deformation over the whole deformation range.

Fe-0,2C-14Mn steel, tensile loaded to 20% elongation. The structure consists of twins of various widths. Note that the twins stop at other twins boundaries. This behavior demonstrates the main difference between twins and plate (twinned) martensite, the latter always terminating at a boundary (see Chapter 5.5).

2.4   FORMATION OF A DISLOCATION SUBSTRUCTURE AT ROOM TEMPERATURE IN METALS OF VERY LOW STACKING FAULT ENERGY

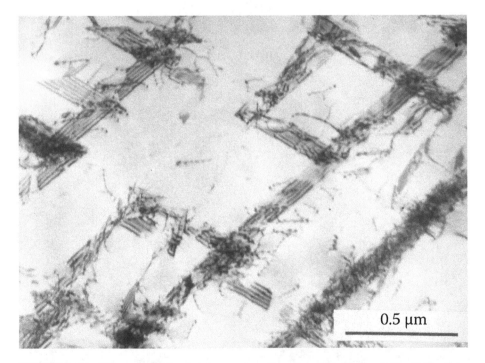

FIGURE 2.16   When the SFE value is very low (less than 10 mJ/m^2), all the dislocations are split into partials separated by wide SF ribbons. The cross-slip of such a complex structure is impossible. It can move only by gliding in its own crystal plane, thus producing a well-defined planar distribution of dislocations. Structures of this type are observed in silver, brass, aluminum bronze, and austenitic steels alloyed with nitrogen.

Austenitic nitrogen steel Fe-18Cr-14Mn-0,6N after 2% cold deformation.

FIGURE 2.17

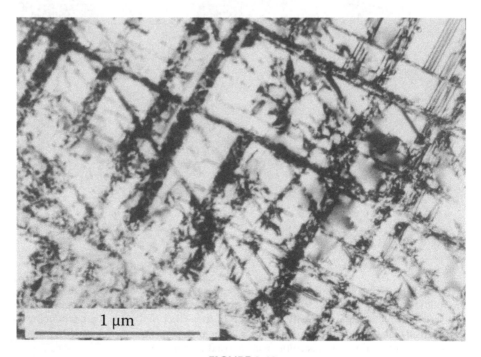

FIGURE 2.18

FIGURE 2.17   Split dislocations in metals of very low SFE usually move in rows, each dislocation following in the track of the preceding one.

Austenitic nitrogen steel Fe-18Cr-14Mn-0,6N after 2% cold deformation. The imaging conditions have allowed only the partial dislocations to be visible, while the SF ribbons are invisible.

FIGURE 2.18   The deformation twinning and the formation of hcp ε-martensite are frequent deformation mechanisms in the fcc crystals of very low SFE. The morphology and the crystallography of both these deformation products are very similar to those of stacking faults, and it is very hard to distinguish them even by TEM.

Austenitic nitrogen steel Fe-18Cr-14Mn-0,6N after 10% cold deformation. The dark bands are probably sheets of ε-martensite because they do not stop when meeting another band but propagate further, suffering only a slight displacement. The ribbons containing black and white fringes are stacking faults.

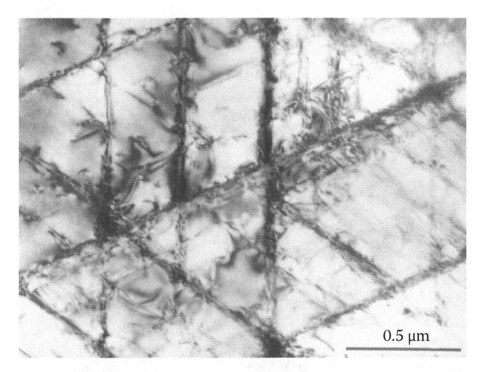

0.5 μm

FIGURE 2.19   The probability of the activation of dislocation movement on a particular close-packed plane depends on the orientation of applied stress to that plane. The movement starts initially only on planes that are "active" with respect to the applied stress. Due to the higher dislocation density, the active planes are visualized in the TEM as dark *slip lines* or *slip bands*.

Austenitic nitrogen steel Fe-18Cr-14Mn-0,6N after 20% cold deformation. Note the 60° angle between the separate bands on the micrograph; this is the angle between the three active {111} planes of the fcc crystal.

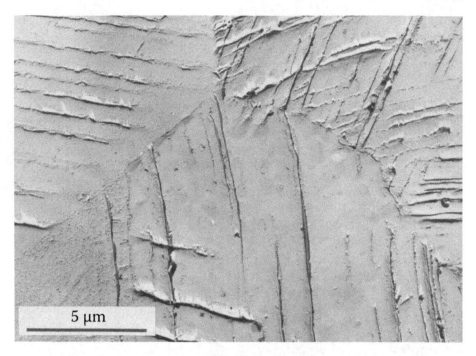

FIGURE 2.20

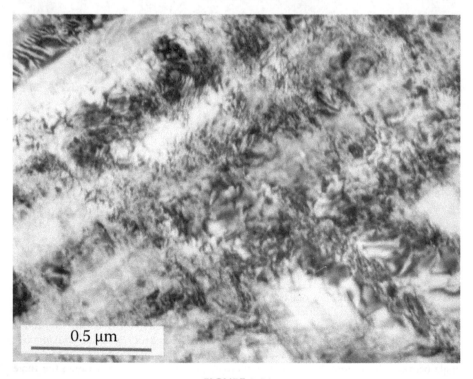

FIGURE 2.21

FIGURE 2.20   The preferred movement of dislocations in separate slip planes results in the formation of more stable *persistent slip bands*. The development of these bands depends on the value of SFE: the lower the SFE, the faster the bands form. The deformation bands are easily detected by light microscopy and by TEM replicas due to the accelerated etching of the bands' intersections with the polished and etched surface of the specimen.

Austenitic nitrogen steel Fe-18Cr-14Mn-0,6N after 20% cold deformation. Replica. The direction of the bands is related to the crystallographic orientation of the active *slip systems* in each specific grain. The bands in adjacent grains have a different direction due to the different crystallographic orientation of the grains.

FIGURE 2.21   The planar dislocation distribution in low-SFE alloys remains unchanged up to a reduction of several tens of percents. With the increase of deformation, the newly introduced dislocations not only join the existing rows but also form new planar rows.

Austenitic nitrogen steel Fe-18Cr-14Mn-0,6N after 20% cold deformation. The number of dislocations in the shown planar rows is so high that it is difficult to resolve the individual dislocations.

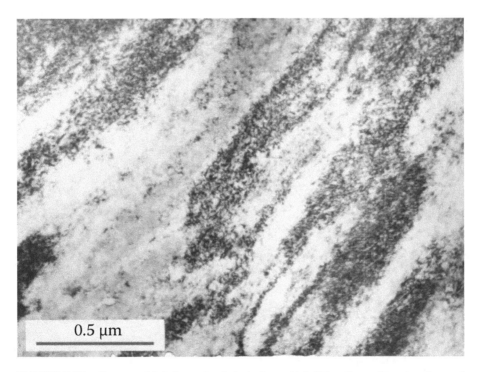

0.5 µm

FIGURE 2.22   Severe cold deformation brings the partial dislocations closer together and forces them to move on all close-packed planes. However, if the SFE value is extremely low, the wide SF ribbons cannot be fully constricted even at deformations above 50%. Thus, severely deformed alloys of extremely low SFE develop a special grid structure—a fine three-dimensional grid of partial dislocations separated by extremely narrow stacking faults.

Austenitic nitrogen steel Fe-18Cr-14Mn-0,6N after 50% cold deformation.

2.5 FORMATION OF A DISLOCATION SUBSTRUCTURE AT HIGHER TEMPERATURE IN METALS OF LOW STACKING FAULT ENERGY

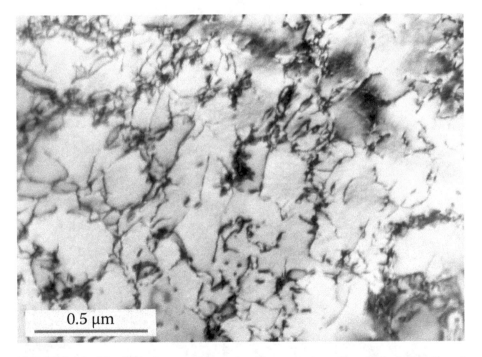

0.5 μm

FIGURE 2.23 The SFE of a metal increases with temperature. The plastic deformation at elevated temperatures ($T < 0.5T_{melt}$—not higher than the temperature of recrystallization) is known in industrial practice as *warm plastic deformation*. The "warm"-deformed medium-low-SFE alloys (copper, austenitic steel Fe-18Cr-9Ni) develop a cell structure similar to that of cold-deformed high-SFE alloys but form a larger number of imperfect cells.

Austenitic steel Fe-18Cr-9Ni after 5% deformation at 400°C.

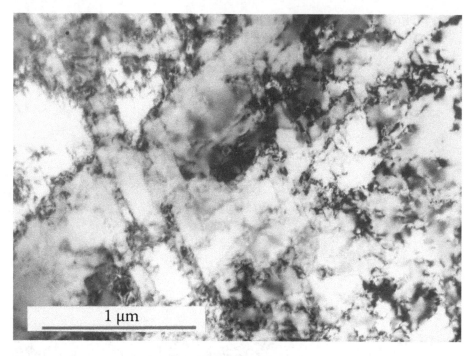

1 μm

FIGURE 2.24

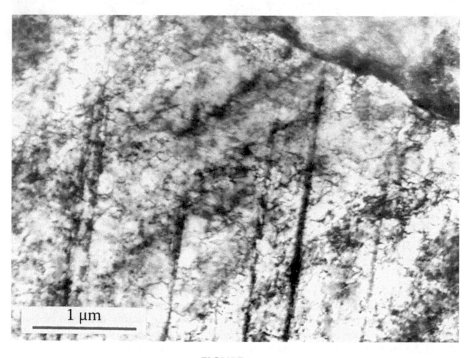

1 μm

FIGURE 2.25

FIGURE 2.24   The increase of SFE in alloys of very low initial SFE (e.g., silver, brass, aluminum bronze, nitrogen-alloyed austenitic steels) at the temperature of warm deformation is not sufficient for the recombination of partial dislocations into unit dislocations. That is why some elements of planar dislocation structure are preserved in these alloys after a small amount of warm deformation.

Austenitic nitrogen steel Fe-18Cr-14Mn-0,6N after 10% deformation at 400°C.

FIGURE 2.25   After a larger amount of warm deformation, the majority of the partial dislocations in low-SFE alloys are brought together to form unit dislocations, which then develop a three-dimensional dislocation distribution. However, the presence of planar bands in the structure indicates that some dislocations still have preserved their separation and have thus continued to move as partials.

Austenitic nitrogen steel Fe-18Cr-14Mn-0,6N after 30% deformation at 400°C.

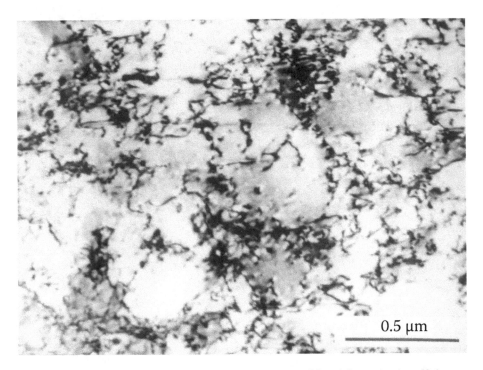

0.5 μm

FIGURE 2.26   The increase of SFE at the temperatures of *hot deformation* is sufficient to make all dislocations become unit ones. These dislocations can move by climb and cross-slip in all the slip planes to form tangles and cells. The cells are similar to the cells of cold-deformed high-SFE alloys except for wider cell walls, the smaller misorientation between cells, and the larger number of dislocations in the cell interior in the hot-deformed low-SFE material.

Austenitic nitrogen steel Fe-18Cr-14Mn-0,6N after 5% deformation at 900°C.

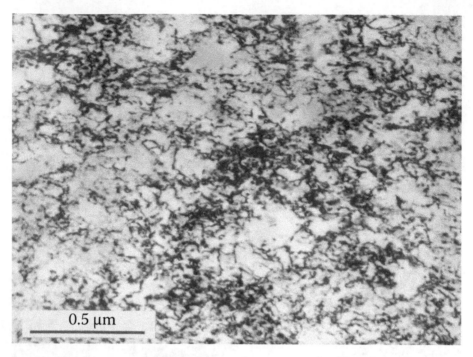

FIGURE 2.27 Even after a large amount of hot deformation, the width of cell walls in low-SFE alloys remains large and the misorientation between cells is small. In this group of alloys, the dislocations never build the substructure typical of the hot-deformed high-SFE alloys (see Figures 2.7–2.9).

Austenitic nitrogen steel Fe-18Cr-14Mn-0,6N after 30% deformation at 900°C.

3 Changes in the Deformation Structure Caused by Heating

A fraction of the energy expended during deformation is stored within the metal as crystal lattice defects, mainly dislocations. At ambient temperature, this causes *strain (work) hardening* of metals. The strain-hardened state of the deformed structure is thermodynamically unstable. The stored energy can be released and the original structure and properties can be partially or completely restored by heating (*annealing*) to elevated temperature through two main *softening mechanisms*: *recovery* and *recrystallization*.

The generalized term *recovery* is used to describe the changes of density and distribution of the deformation-induced defects of the crystal structure by heating at temperatures lower than the temperature of recrystallization, that is, before the appearance of new recrystallized grains. The stored energy is lowered by two principal recovery mechanisms:

- Annihilation of point defects (mainly excess vacancies) and dislocations of opposite sign
- Rearrangement of the dislocations into lower-energy configurations

The process of formation and migration of *low-angle boundaries* in the deformed metal during recovery is called *polygonization*. The enhanced annihilation and recombination of dislocations, assisted by increased temperature, results in a reduced number of dislocations in cell walls and a larger misorientation across the walls. Thus, the cell walls undergo a gradual transformation into low-angle boundaries.

The term *recrystallization* includes all structural processes involving migration of high-angle boundaries and the growth of new strain-free grains in the deformed structure. It results in an entirely changed microstructure: reduced dislocation density, strain-free grains with different size and shape, recrystallization texture, as well as in complete restoration of properties.

When heating a cold-deformed metal or alloy at temperatures higher than the recrystallization temperature, a *primary recrystallization* takes place. It starts by nucleation of new crystals of low internal energy (recrystallization nuclei) separated by high-angle boundaries from the surrounding deformed metal. It then proceeds by growth of the nuclei until new grains completely replace the deformed or recovered structure. In single-phase structures, the new grains nucleate in the regions of maximum accumulated energy, that is, the regions of maximum quantity of defects and lattice distortions. The grains can also nucleate at grain boundaries, large subgrains, deformation bands, or at particles of second phase, which have precipitated before or during the heating *(particle-stimulated nucleation)*.

59

Since the recovery and recrystallization are driven by the reduction in the stored deformation energy, a relationship exists between both processes. Enhanced recovery may slow down the recrystallization by lowering the stored energy. On the other hand, if recrystallization has taken place in the deformed structure, no further recovery is possible. This "competition" between the two processes depends on the deformation temperature and ratio, the SFE, and the recrystallization temperature of the metal.

Grain growth is a process that involves migration of the high-angle boundaries of the recrystallized grains. It results in a *normal grain growth*. This process should not be confused with *secondary recrystallization* or *abnormal grain growth*, also known as coarsening, which leads to discontinuous growth of separate very large grains into the rest of the structure.

Hot deformation promotes recovery processes because they rely on thermally activated mechanisms. Two processes of "self-annealing" (metal softening) that can occur during deformation at elevated temperatures are *dynamic recovery* or *dynamic polygonization* (most distinctly seen in high-SFE alloys) and *dynamic recrystallization*. Which one of the specific mechanisms of dynamic polygonization will operate—formation of subgrains, growth of large subgrains by coalescence, rotation of subgrains—depends on both the temperature and the strain rate. The determining factors for the occurrence of dynamic recrystallization and the extent of dynamic softening are the nature of material (alloy composition and SFE value), the temperature, the amount of deformation, and the strain rate. Dynamic recrystallization can occur in commercial alloys such as austenitic-alloyed steels, heat-resistant nickel-based alloys, and brass during sufficiently high hot working.

FIGURE 3.1

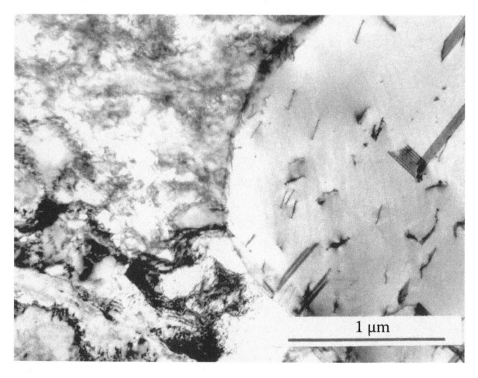

FIGURE 3.2

FIGURE 3.1   Annealing of cold-deformed metals results in the decrease of the number of dislocations in the cell's interior while the dislocation tangles in the cell walls change into more regular dislocation networks. The walls gradually transform into subboundaries, and the cells into subgrains free of dislocations. These microstructural features are the main indicators that polygonization has occurred.

Aluminum after 30% cold deformation followed by 5 min annealing at 200°C.

FIGURE 3.2   If the amount of cold deformation is low or the duration of annealing is short, recrystallization can occur only in some areas of the metal where the level of accumulated energy is higher. The new recrystallized grains of this *partially recrystallized structure* are surrounded by a deformed matrix.

Austenitic nitrogen steel Fe-18Cr-14Mn-0,6N after 50% cold rolling, followed by 30 min annealing at 900°C.

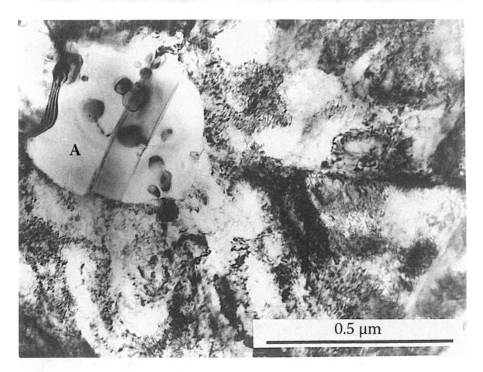

A

0.5 μm

FIGURE 3.3   The nucleation of new grains very often occurs on particles of a second phase that existed in the deformed structure prior to annealing or that had precipitated before the start of recrystallization.

Austenitic nitrogen steel Fe-18Cr-14Mn-0,6N after 50% cold rolling, followed by 2 min annealing at 550°C. In the present case, recrystallization has started preferentially only in area A, where the first Cr_2N particles had precipitated, while no precipitation can be seen in the rest of the austenite, where the deformation structure remains unchanged. The incubation period for recrystallization was clearly longer than that for nitride precipitation.

FIGURE 3.4

FIGURE 3.5

FIGURE 3.4 Recrystallization kinetics depends on both the size and the density of particles present in the material before annealing. Second-phase particles can serve as recrystallization centers (nuclei for new grains) only if their size exceeds some critical value specific to the given cold work reduction.

Austenitic nitrogen steel Fe-18Cr-14Mn-0,6N after 50% cold rolling, followed by 5 min annealing at 550°C. In this micrograph the Cr_2N particles have precipitated throughout the metal, but their size has reached the critical value only in the central part of the image where, as a result, recrystallization has started first. New grains of smaller size have nucleated later on the smaller particles. Polygonization commenced in the rest of the material in which no precipitation has occurred.

FIGURE 3.5 After nucleation, new recrystallized grains grow through migration of high-angle grain boundaries, a process driven by the remaining stored energy. When the annealing temperature is proper (different for the different alloys and dependent on the amount of prior deformation) and the annealing time is sufficient, the new grains totally consume the old deformed structure until the primary recrystallization is complete.

Austenitic nitrogen steel Fe-18Cr-14Mn-0,6N after 50% cold rolling, followed by 30 min annealing at 550°C. In this micrograph the primary recrystallization has finished, the size and shape of the fully recrystallized grains are now totally changed, and the structure is strain-free, with very low dislocation density.

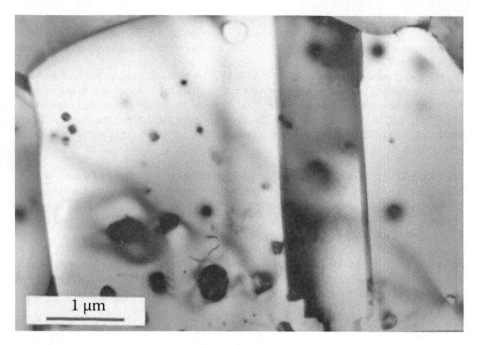

FIGURE 3.6

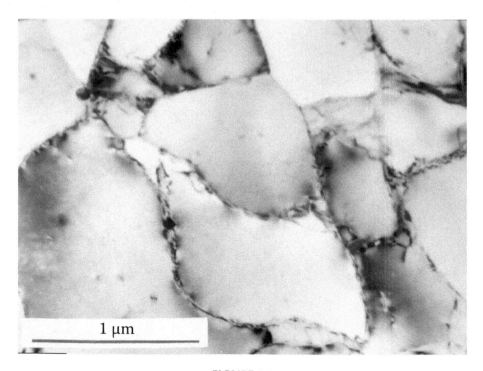

FIGURE 3.7

FIGURE 3.6   Keeping the primary recrystallized metal at a high temperature for a prolonged period of time results in normal growth of the new grains by migration of the highly mobile large-angle boundaries. It results in continuous increase of the mean grain diameter and reduction of the total number of grains. However, the crystallographic orientation and the location of some of the new grains in the structure appear to be more favorable for their growth. These grains can then grow to a larger size at the expense of their smaller-sized neighbors.

Austenitic nitrogen steel Fe-18Cr-14Mn-0,6N after 50% cold rolling, followed by 2 h annealing at 550°C. Note the annealing twin in the middle of the large grain shown; it is typical for the recrystallized structure of low-SFE metals.

FIGURE 3.7   The hot deformation of many high-SFE metals is usually accompanied by dynamic recovery. For example, during extrusion, forging, or hot rolling in commercially pure aluminum and aluminum alloys, a dynamic polygonization takes place.

Aluminum after 20% extrusion at 300°C. The polygonized substructure of the hot-deformed semifinished product is much more stable than the structure, produced by cold deformation.

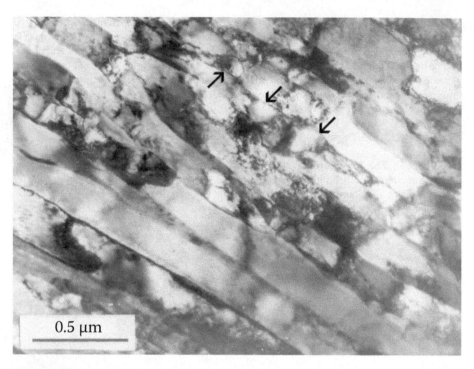

0.5 μm

FIGURE 3.8 Dynamic recrystallization can occur in some alloys during severe, high-speed cold deformation. The signs of dynamic recrystallization can be observed only if the deformed material is rapidly cooled to prevent static recrystallization resulting from the heat generated by the intensive deformation process.

Copper after 75% cold extrusion. The small final cross-section of the extruded rod ensured rapid cooling. As a consequence, several dynamically recrystallized grains (shown by arrows) can be observed in the final structure.

4 Growth of the Crystals and Rapid Solidification

4.1 GROWTH OF THE CRYSTALS

There are three well-established mechanisms behind the growth of crystals in pure solids: *continuous, lateral,* and *spiral growth.* All three are related to the type of solid/liquid interface and its migration.

The term *continuous growth* originates from the mode of propagation of the diffuse disordered solid/liquid interface, normally to itself, by joining atoms arriving from the liquid at random positions everywhere over the solid surface. This mechanism is characteristic for metallic systems.

In the case of atomic smooth interfaces, characteristic mostly of nonmetallic materials, the *lateral growth* process takes place. This is facilitated by the presence of specific *ledges* and *jogs* and partly filled interface layers. These are preferred sites for the atoms to attach themselves to the solid because of the lower or no increase in interfacial energy compared to that associated with a perfectly flat surface.

The other mechanism affecting crystal growth is *screw-dislocations activity.* The well-known model of *spiral growth,* which has been observed and proven in Mg, Co, Ag, and so on, explains why the experimentally observed growth rate is much higher than the rate predicted by the surface repeated nucleation theory. The exit of screw dislocation at the crystal surface creates a *permanent step (ledge)* of atomic dimensions. When the atoms continuously arrive and attach at equal rate to the step, they make the step rotate around the exit point of the dislocation, but they never remove it. The decrease of angular velocity of rotation away from the dislocation core causes the gradual transformation of the ledge into a permanent spiral, each winding of which lifts the solid surface by one interatomic distance.

The direct observation of crystal growth during the solidification of metals is complicated (if not impossible), because metals as well as many nonmetals are opaque. For this reason, the processes are modeled using a variety of transparent materials and saturated aqua solutions of salts. The phenomena connected with the growth of crystals can be studied in the TEM using a preparation technique called *gold decoration*—vacuum evaporation and deposition of gold on heated fresh cleaved crystals, usually NaCl. Gold particles preferentially attached to the ledges and jogs facilitate the observation of the monatomic steps on the surface.

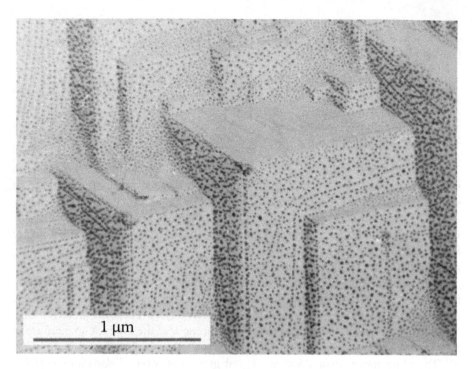

FIGURE 4.1

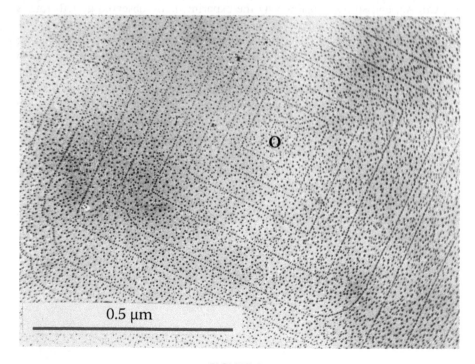

FIGURE 4.2

FIGURE 4.1 *The replica of a cleaved NaCl single crystal shown in this micrograph was prepared by the gold decoration technique.* The cleavage has followed the crystal planes of the NaCl cubic crystal lattice and thus the growing ledges. The deposited gold particles subsequently make the ledges clearly visible.

FIGURE 4.2 *A screw dislocation, which terminated during the growth of the NaCl crystal, has created a ledge on the surface fixed at the exit point O.* As the growth proceeded and atoms continuously added to the step, it transformed into a growth spiral. Each winding of the spiral raised the surface of the growing crystal by one interatomic distance.

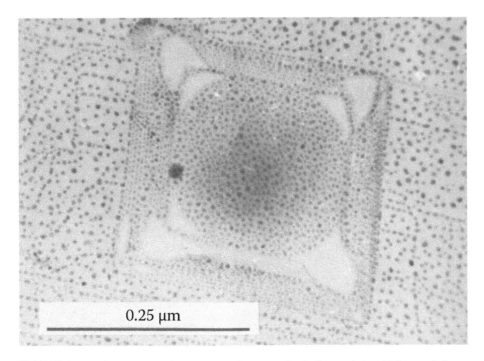

0.25 µm

FIGURE 4.3 *This micrograph illustrates the case of spiral growth in which two dislocations of opposite sign participate.* The growth steps form a closed planar terrace, which is a part of a conical growth pyramid.

4.2 RAPID SOLIDIFICATION PROCESS

A rapid solidification process (RSP) requires cooling rates typically in the range 10^4–10^7 Ks^{-1}. In practice, different techniques can be utilized to achieve RSP, such as pouring a fine stream of liquid metal onto a rotating chilled cylinder or spraying liquid microdrops onto a cooled (usually copper) substrate. The technique of rapid quenching from a melt of a desired chemical composition is termed *melt spinning*.

The high cooling rate gives undercooling that results in a number of unusual structural effects: a glassy (amorphous) state, refinement of the grains, extended solute solubility in solid state, formation of metastable crystal phases, and lowering of martensitic transformation temperature. These effects are used for the synthesis of new categories of metallic materials, namely, the *amorphous metals* usually referred to as *glassy metals* or *metallic glasses* and the *rapidly solidified microcrystalline alloys*.

Glassy metals are actually multicomponent alloys, produced by RSP, which preserve the short-range order specific to the liquid metal in the solid state. The amorphous, noncrystalline structure ensures unusual and potentially useful physical (magnetic, electrical) and mechanical properties as well as yielding properties that are highly isotropic. Amorphous structure does not contain dislocations and grain boundaries, so the metallic glasses have a high strength and resistance to plastic deformation, wear, and corrosion. The formation of amorphous structure requires a suitable chemical composition (components acting as glass formers) and very fast cooling; the latter limits the maximum thickness of the produced metallic glasses.

Microcrystalline alloys produced by RSP have many advantages compared to conventional alloys, such as improved mechanical properties, reduction or elimination of micro-segregation, and increased solubility in the solid state, which consequently provides a greater strengthening effect on aging. The ultra-fine-grained structure exerts a marked effect on strength and plasticity while serving as a basic microstructure for superplastic deformation.

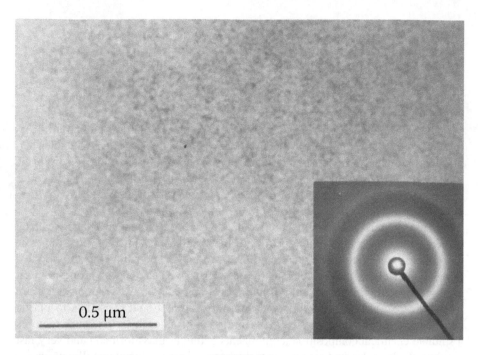

FIGURE 4.4

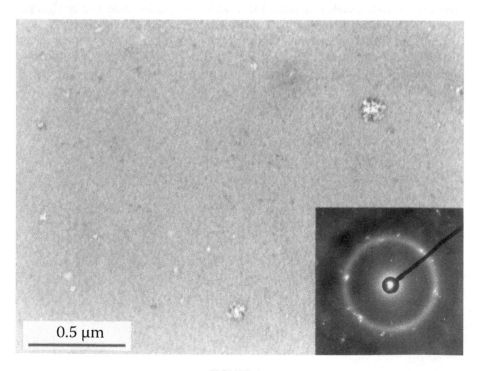

FIGURE 4.5

FIGURE 4.4 An amorphous (glassy) state can be obtained in some metal/metal or metal (Fe, Ni, Co, Mo, Pd)/nonmetal (B, Si, P) systems by undercooling at cooling rates equal to or greater than 10^5 Ks^{-1} to suppress the normal crystallization process.

Amorphous Ni-Si-B alloy, melt spun at cooling rate of 10^5 Ks^{-1}. The presence of diffuse rings and the absence of diffraction spots in the electron diffraction pattern are an indication of the absence of a crystalline structure, that is, the presence of an amorphous state; see inset to the figure.

FIGURE 4.5 The amorphous (glassy) solid state of alloys is metastable. When such an alloy is heated above some specific temperature for that material, the alloy undergoes a spontaneous crystallization process similar to the crystallization of a molten metal that has been undercooled below its freezing temperature. The nucleation starts at random positions in the amorphous alloy with the formation of single crystals and continues by their growth, which takes on a dendritic form.

Single crystal nuclei in an amorphous Ni-Si-B alloy after heating the alloy for 5 min at 700°C. The appearance of the crystal nuclei immediately changes the nature of the electron diffraction pattern. The newly formed crystal planes produce diffraction spots (reflections) in addition to the diffuse rings obtained from the surrounding amorphous material; see inset to the figure.

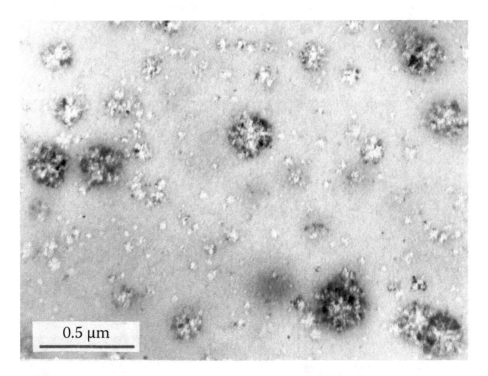

0.5 μm

FIGURE 4.6 Crystallization continues during the isothermal heating of a glassy metal by the growth of the existing crystal nuclei and by the formation and growth of new crystal nuclei. Consequently, the number of diffraction spots in the diffraction pattern increases and they start to form rings, typical for the polycrystalline structures.

Crystal nuclei in an amorphous Ni-Si-B alloy after heating the alloy for 10 min at 700°C.

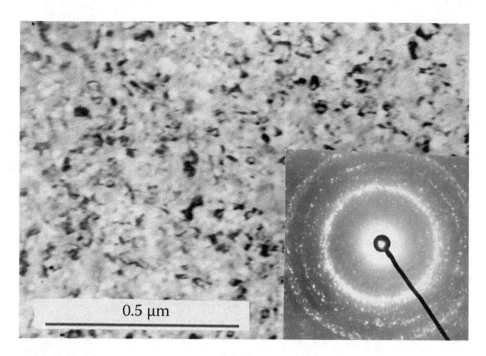

0.5 μm

FIGURE 4.7

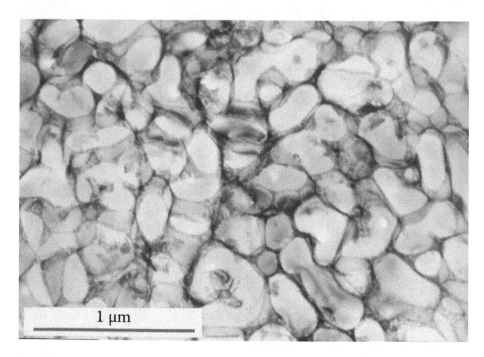

1 μm

FIGURE 4.8

FIGURE 4.7 The cooling of a liquid metal or alloy at lower rates does not allow an amorphous state to be established but can produce an ultra-fine microcrystalline structure through the simultaneous nucleation of crystals throughout the entire volume of the material.

Microcrystalline Ni-Si-B alloy that was obtained by cooling at a rate of 10^4 Ks^{-1}. The rapid quenching has produced randomly oriented microcrystals of a size less than 1 micrometer, which resulted in complete rings of diffraction spots formed from all the grains; see inset to the figure.

FIGURE 4.8 Even when cooled very rapidly, some molten alloys do not freeze in the amorphous state. Instead, they develop a microcrystalline structure with an unusually high supersaturated solid solution, refined grains, and modified morphology of the precipitated phases.

Al-11%Si alloy, quenched from melt on cooled copper substrate. The same alloy conventionally cooled shows a mean diameter of α_{Al}-grains of several tens of micrometers. The rapid cooling refines the grains to several tens of nanometers.

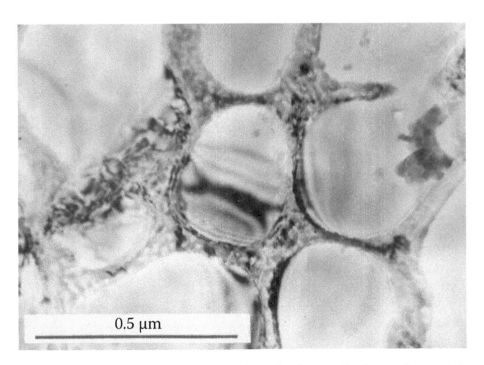

0.5 μm

FIGURE 4.9 The rapid solidification of an AlSi alloy dramatically changes the morphology of aluminum-silicon eutectic that borders the ultra-fine grains, even compared to the fine fibrous eutectic in modified alloys.

Al-11%Si alloy, quenched from melt on cooled copper substrate. The size of the rounded silicon particles is reduced to less than several nanometers.

5 Solid-State Phase Transformations

Solid-state phase transformations play an important role in the development of the structure and properties of metals and alloys. Polymorphic transformations and precipitation reactions are the main types of phase transformations that occur in solids. Well-known examples are the polymorphic transformations of iron, titanium, and cobalt; the hardening of materials by martensitic transformations; and the strengthening (dispersion hardening) by solid-state precipitation. The driving force behind all structural and phase transformations from their start to their completion at a given temperature and pressure is the decrease of the Gibbs free energy. The knowledge of phase transformation mechanisms provides the basis for the theory and practice of the heat treatment and the processing of metals.

According to the classification system based on the physical mechanisms of the reaction and the mode of interphase migration, solid-state phase transformations are divided into two main groups:

- *Diffusional transformations*. These transformations take place by thermally activated atom movement and require diffusion either through the lattice or across a *nonglissile interface*. These transformations are also termed *civilian* due to the uncoordinated atomic movements. The composition of the parent and the product phases may be the same or different. The majority of phase transformations (phase precipitation, eutectoid and massive transformations, ordering) are diffusion controlled.
- *Diffusionless transformations*. No diffusion is involved in these athermal transformations, which are also classified as *military* because of the coordinated movement of the atoms. The phase interface, which must be either coherent or semicoherent, is glissile, and the individual atoms move a distance shorter than one interatomic spacing without changing the neighborhood of the nearest atoms. The product phases preserve the chemical composition of the parent phase. A typical representative of this type of transformation is the martensitic transformation, which is achieved by a shear mechanism.

There are *intermediate transformations* that simultaneously possess the characteristics of both diffusional and nondiffusional reactions. The bainite transformation, which takes place by both a shear mechanism and long-range diffusion, is referred to as intermediate between the martensitic and the pearlitic transformations.

The diffusional processes of phase precipitation are classified as *continuous* or *general* and *discontinuous* or *cellular* depending on the mechanism of the nucleation and growth of the new phase. In a continuous precipitation, the entire matrix is continuously and uniformly depleted of solute elements, thus producing a single precipitate phase with a random distribution. The discontinuous precipitation starts at grain boundaries or phase interfaces and produces a two-phased structure with a cell morphology.

5.1 CONTINUOUS PRECIPITATION IN AGE-HARDENING ALLOYS

The basic requirement for a precipitation-hardening alloy system is that the solid solubility limit should decrease with decreasing temperature. The precipitation occurs when the alloy, solution treated and quenched to obtain a supersaturated solid solution (SSSS), is allowed to "age" for a sufficient period of time at temperatures below its solvus temperature. The precipitation process is accompanied by changes in the properties of the material. Thus, aging (precipitation hardening) has been used for many years in the commercial heat treatment of numerous engineering alloys.

The precipitation sequence in most age-hardening alloys is complex and multi-stage and usually involves formation of clusters, zones, and several transition phases. Commonly the precipitation reaction is written as SSSS → clusters → GP-zones → $\beta'' \to \beta' \to \beta$, where β'' and β' are transition phases and β is the equilibrium phase.

Small *clusters of solute atoms* form in the lattice in the earliest stages of the precipitation sequence. The clusters develop into *Guinier–Preston zones (GP-zones)*— solute-enriched areas with atomic arrangement fully identical with the crystal lattice of the matrix. The nucleation of clusters (zones) and growth of zones occur homogeneously throughout the supersaturated matrix. The presence of excess vacancies in the structure is decisive for the zones' formation because vacancies provide fast diffusion for solute atoms in the matrix lattice.

Precipitation via *transition (intermediate, metastable) phases* provides a more rapid decrease of the SSSS free energy than by direct transformation to the equilibrium phase. When the SSSS decomposes via transition phases, the activation energy barrier corresponding to each subsequent stage of their precipitation is much smaller in comparison to the barrier associated with the direct precipitation of the equilibrium phase. These phases nucleate heterogeneously on preferred sites such as zones or crystal imperfections because of the lower nucleation energy in the regions of the disturbed atom arrangement. The ability of crystal imperfections to assist nucleation increases in the same order as the energy of the defects associated with the atomic disturbance in their vicinity: vacancy clusters, dislocations, SFs, subgrain boundaries, and grain boundaries.

Continuous precipitation can be microscopically *uniform* or *homogeneous,* when the precipitates are uniformly distributed in the entire volume of the grain. In contrast, *localized* or *heterogeneous,* precipitation occurs when precipitates nucleate heterogeneously on some preferred sites. The latter precipitates are coarse and obtain uneven distribution.

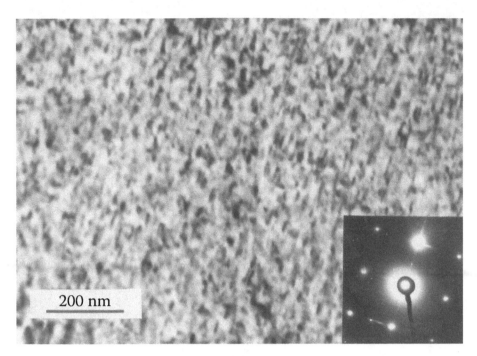

FIGURE 5.1   The number of transition phases in an alloy system depends on the ratio of the solid solution supersaturation to equilibrium phase content. Several intermediate precipitation stages are observed in the Al-Cu system. The first stage for an Al-4%Cu alloy is the homogeneous nucleation of copper-rich zones, which are termed *GP-1 zones*. These zones typically are considered to be discs of copper atoms on the {100} planes of the aluminum matrix. Due to their small size, they cannot be directly observed in routine TEMs. The distortion of the crystal lattice around the zone caused by the atomic size difference between the copper and aluminum atoms is very small and thus can be accommodated solely by elastic strains. The only indication for the presence of the zones is the specific roughness of the foil surface, the coherency misfit strain contrast, and the streaking of diffraction spots in a direction normal to {100}$_{Al}$ (see inset). Both the strain contrast and the streaking are derived from the elastic distortion of the crystal lattice in a direction perpendicular to the plane of the zones.

Al-4%Cu, aged to produce GP-1 zones.

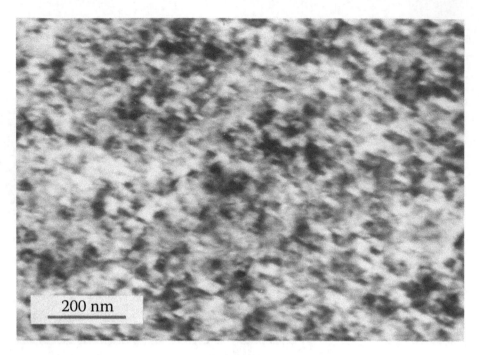

FIGURE 5.2

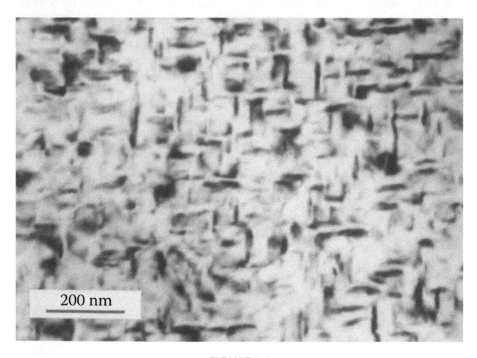

FIGURE 5.3

FIGURE 5.2 The GP-zone stage is characteristic of the precipitation sequence in many engineering aluminum-based alloys, copper-beryllium, and so on. The shape of zones depends on the difference between the atomic diameters of solid solution constituents and can be spherical (Al-Ag, Al-Zn, Al-Zn-Mg), rodlike (Al-Cu-Mg), needle-shaped (Al-Mg-Si), or disc-shaped (Al-Cu, Cu-Be). The size of the zones for each alloy depends on the aging time and temperature.

Al-4%Cu, aged to produce GP-1 zones. The disc-shaped zones lie on $\{111\}_{Al}$ planes.

FIGURE 5.3 The misfit between the atomic arrangement in the zone and the surrounding matrix increases with aging time due to the increasing content of solute atoms in the zone. The term *zone precipitation* is related mainly to the well-studied aluminum alloys. It applies to the aging process up to the stage when the misfit becomes too large to be compensated solely by elastic strains. In the Al-4%Cu alloy, the zone stage includes formation of *GP-2 zones,* also referred as the *metastable θ″-phase.* When the diameter of zones reaches about 10 nm, the atomic arrangement of GP-1 transforms into GP-2, which is fully coherent with the fcc lattice of aluminum. θ″ precipitates are larger than the GP-1 zones, being up to ~10 nm thick and 100 nm in diameter.

Al-4%Cu alloy, aged to produce GP-2 zones (metastable θ″-phase). The shadows around the coherent θ″-phase are due to the coherency strain fields between the zone and the matrix.

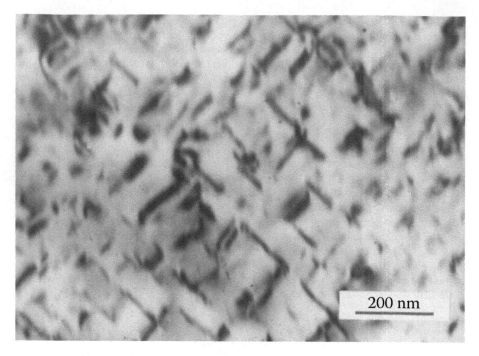

200 nm

FIGURE 5.4

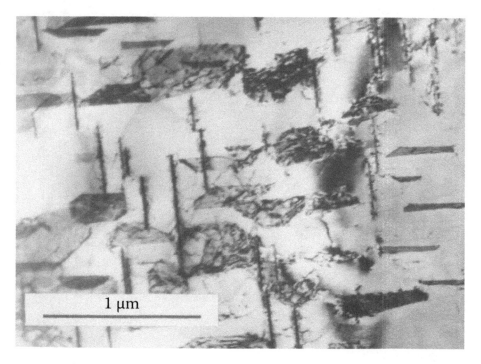

1 μm

FIGURE 5.5

FIGURE 5.4   Upon aging, the fully coherent θ''-precipitates (GP-2 zones) become larger and the atomic misfit in some of the crystallographic directions increases. When the size of the precipitates exceeds some critical dimension, the coherency strains and misfit at the matrix/precipitate interface become so large that they have to be compensated by the creation of new phase interfaces containing *misfit dislocations*. If the interfaces in the different crystallographic directions are of mixed type, coherent and noncoherent, the precipitate is defined as being *semicoherent*. The semicoherent θ' phase, with a tetragonal crystal structure, forms continuously from the GP-2 zones by an *in situ* transformation because the structures of these phases are similar.

 Al-4%Cu alloy, aged to produce semicoherent θ''-phase. The dark shadows around the semicoherent θ'-particles are weaker than the shadows around the coherent θ''-phase (see Figure 5.3) because the latter has more strain associated with it. The interface misfit dislocations around the θ'-particles cannot be resolved.

FIGURE 5.5   Equilibrium phases can be differentiated from the matrix when the number of solute atoms in the enriched zones reaches the concentration of the phase equilibrium of the alloy system. The equilibrium phase in Al-4% Cu alloy is the intermetallic compound $CuAl_2$—a noncoherent body-centered tetragonal θ-phase.

 Al-4%Cu alloy, aged at high temperature to produce equilibrium θ-phase ($CuAl_2$) and semicoherent θ'-phase. Notice the absence of strain fields (dark shadows) around the plate-shaped equilibrium θ-particles and the presence of interfacial dislocations accommodating the misfit between the semicoherent θ'-phase and the matrix.

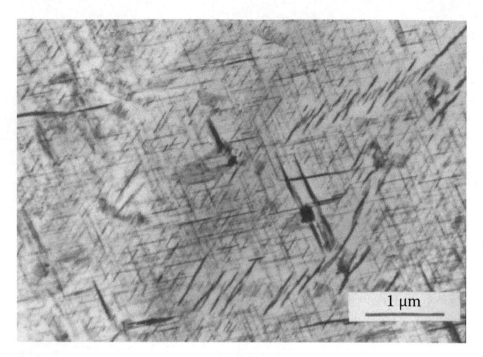

FIGURE 5.6

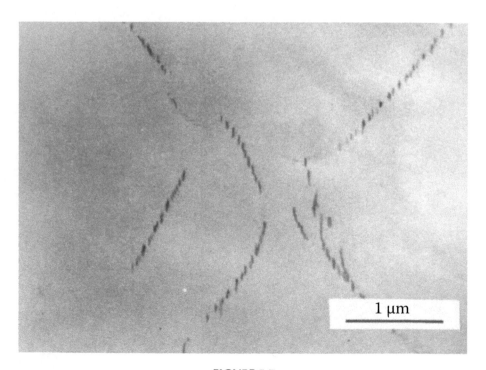

FIGURE 5.7

FIGURE 5.6 In many alloys, the nucleation and growth of phases occurs on preferred matrix planes, called *habit planes,* and under a specific crystallographic orientation relationship with some of the matrix crystal planes or directions. Under these conditions, the new phase usually develops a planar interface with the matrix and takes on a regular geometric shape. In the case of Al-4 Cu, the $\{100\}_{Al}$ is the habit plane of the GP-1 zones, of the metastable θ'' (GP-2) and θ' phases, and of the equilibrium $CuAl_2$ phase. This results in the specific morphology, which is referred to as a *Widmanstätten structure.*

Al-4%Cu alloy aged to produce equilibrium θ-phase.

FIGURE 5.7 The nucleation of a second phase at structural irregularities is preferred because a smaller amount of energy is required to create a nucleus of critical size in their vicinity. The heterogeneous nucleation on crystal defects proceeds faster than the homogeneous nucleation in the perfect matrix, but the general decomposition rate is limited by the number of structural irregularities.

Cast alloy Al-6%Cu-0,2%Ti. In this micrograph, heterogeneous precipitation has started at dislocations earlier than homogeneous precipitation in the matrix. The θ-particles decorating the dislocation lines have grown up to 150 nm in size, and still there is no indication of any decomposition in the rest of the matrix.

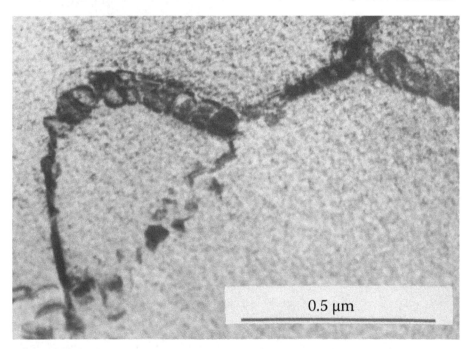

0.5 μm

FIGURE 5.8 There are several reasons for dislocations assisting nucleation: (a) the strain field around dislocation lines reduces the total strain energy of the nuclei; (b) the solutes segregated on dislocation lines, the Cottrell atmospheres around dislocations, and the Suzuki atmospheres on the SFs provide enriched sites in the matrix; and (c) the dislocation lines provide fast diffusion paths along the dislocation pipe-shaped core.

Cast alloy Al-6%Cu-0,2%Ti after low-temperature aging. The precipitates of θ'-phase that heterogeneously nucleated on the dislocation lines grow faster than those in the matrix. The contrast in the rest of the volume indicates that the decomposition in the perfect matrix is still at the GP-zones stage.

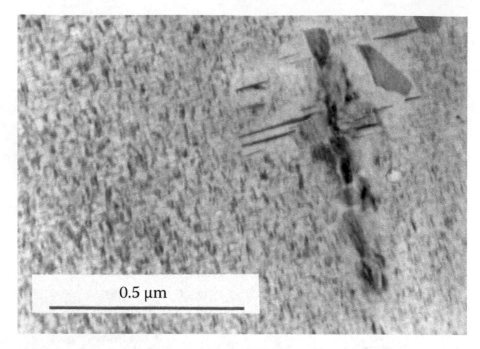

0.5 μm

FIGURE 5.9

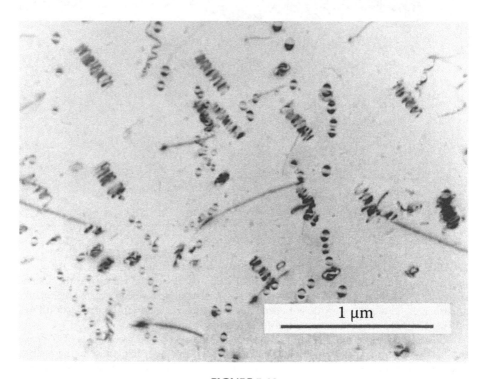

1 μm

FIGURE 5.10

FIGURE 5.9   The accelerated growth of heterogeneously precipitated phases causes solute depletion at some distance from the particle, which retards the homogeneous decomposition in the surrounding matrix.

Cast alloy Al-6%Cu-0,2%Ti after low-temperature aging. Note the Cu-depleted halo around the heterogeneously precipitated phases and the absence of such a zone around the θ-particles.

FIGURE 5.10   As in the case of homogeneous precipitation, excess vacancies assist the heterogeneous nucleation. The solute atoms are attracted to the dislocation loops that border the complexes of vacancies and to the helical dislocations produced by the interaction of vacancies with screw dislocations. This mechanism of vacancy-assisted heterogeneous precipitation is often observed in aluminum alloys, which are especially susceptible to the quenching-in of excess vacancies.

Al-6%Cu–2%Zn alloy after quenching and subsequent aging. Solute atoms have been absorbed onto the helical dislocations left by quenching and have formed disc-shaped precipitates on each spiral of the helicoids.

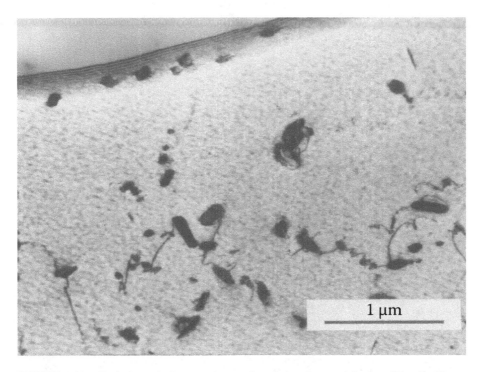

1 μm

FIGURE 5.11   Grain boundaries are also preferred sites for precipitation. The significant amount of energy that is associated with them can be reduced by boundary segregation of solute atoms, followed by precipitate nucleation and growth. The susceptibility to grain boundary segregation increases with the interfacial energy, that is, with the magnitude of the misorientation across the boundary.

Al-6%Cu-0,2%Ti after quenching and a low-temperature aging. The size of the particles on the grain boundary and on the dislocation lines has reached several hundreds of nanometers, but the decomposition in the matrix is still at the zone stage. This shows once more that the heterogeneous precipitation starts much earlier than the homogeneous precipitation in the matrix.

FIGURE 5.12

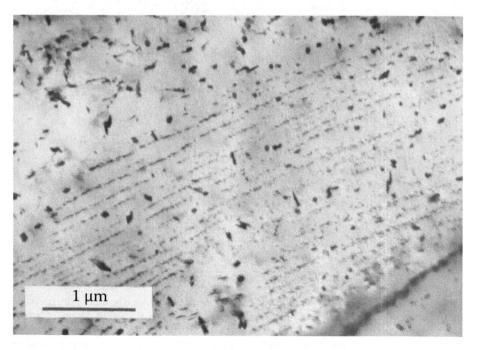

FIGURE 5.13

FIGURE 5.12   Grain boundaries are known to be effective vacancy sinks, and this results in the formation of vacancy-depleted zones around them. On the other hand, the solute atoms from the grain boundary areas are often consumed for heterogeneous boundary precipitation. The more difficult nucleation and growth of phases in the solute-depleted regions causes the formation of a *precipitate-free zone (PFZ)* around such boundaries. PFZs in age-hardened alloys are unwanted due to their detrimental effect on the mechanical properties and corrosion resistance.

Al-6%Cu-0,2%Ti after quenching and a short high-temperature aging. The relatively low rate of cooling from the solid solution temperature for this alloy has promoted grain boundary precipitation and caused the formation of PFZs that are about 1.5 μm wide. Fast cooling during quenching plus prolonged aging at a lower temperature can help to reduce the width of the PFZ.

FIGURE 5.13   Segregation of impurity atoms and precipitation on grain boundaries are often observed at high temperatures. The dragging of boundaries "decorated" with segregated atoms or particles requires too much energy. The only possible mechanism for the boundary to migrate is to bow out between the "decorations" and then to tear away from them. The unrestrained boundary can then move freely only a short distance until new atoms and particles decorate it and stop it. This process can be repeated many times. The precipitates, which have been left behind by the released boundary, indicate the successive positions of boundary during its migration.

Vanadium-alloyed carbon steel after normalization. The austenite-ferrite boundary has been locked by carbon atoms during the polymorphous transformation. Vanadium carbonitrides have had sufficient time to precipitate as a chain along the immobile boundary. The parallel carbide rows seen in the micrograph indicate the successive positions at which the boundary has repeatedly released itself from the precipitates to continue its migration.

5.2 INTERACTION OF DISLOCATIONS WITH SECOND-PHASE PRECIPITATES

Precipitation hardening involves the strengthening of alloys resulting from interaction of dislocations with those precipitates, formed during aging, which are effective obstacles for their free movement. The amount of strengthening achieved by aging of commercial alloys is determined to a large extent by the nature, type, properties, and morphology (size, shape, density, and distribution) of precipitates. Precipitate particles can impede the dislocation motion through a variety of interaction mechanisms, some of which can operate simultaneously. The model of *internal strain hardening* distinguishes two main cases depending on the distribution of the obstacles: (a) the stress fields are closely spaced and the dislocation line has to *curl* between them, or (b) the fields are widely spaced and the dislocation line has to *pass around* them, subsequently leaving a *dislocation loop* around each of them. The change of solvent and solute/neighbor atom arrangement produced by the cutting (shearing) action of the dislocation as it passes through a zone or a coherent precipitate causes *chemical hardening*. The model of *dispersion hardening* applies to the later aging stages when the precipitates are incoherent, they cannot be cut by the dislocation or deformed with the rest of the matrix, and an increase of applied stress is required for the dislocation line to bypass the particles, leaving dislocation loops around them.

The direct study of dislocation–precipitate interactions is rather difficult because the processes occur at the matrix/precipitate interfaces and sometimes are rather difficult to be resolved.

FIGURE 5.14

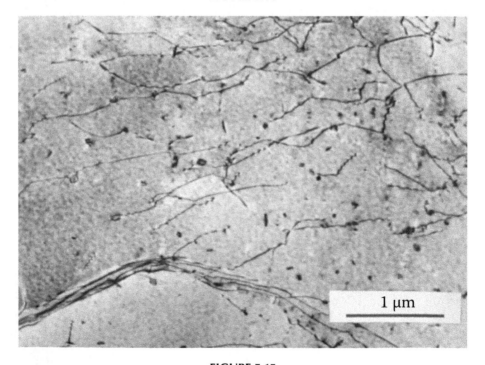

FIGURE 5.15

FIGURE 5.14 Aging treatment, producing a high density of GP-zones or coherent precipitates, results in substantial strengthening because the stress fields around them are effective obstacles to dislocation movement. In Al-4%Cu alloys, GP-2 zones (θ'') are considered to be responsible for the peak strength effect.

Al-4%Cu alloy aged to the GP-zone stage. The internal strain-hardening model is demonstrated in this micrograph by the dislocation lines that have to bow around the zones due to their strain fields.

FIGURE 5.15 The simultaneous formation of zones and precipitation of semicoherent particles guarantees the peak mechanical strength of the alloys because the effect of the strain fields of the zones is superimposed on the effect of the strain fields of the particles.

Al-7%Zn-2%Mg-1%Cu alloy, mold cast. In this micrograph, the dislocation lines have curled due to their interaction with the strain field of the zones while also being additionally "pinned" by the semicoherent particles of the complex intermetallic phases.

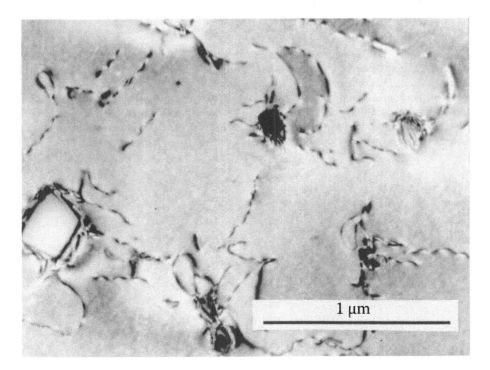

FIGURE 5.16 When the external stress increases and the bowing of the dislocation line becomes large, the dislocation can tear away from the obstacle, leaving a dislocation loop around it.

Al-4%Cu alloy aged to produce θ-phase precipitation. The dislocation loops and the dark shadows produced by the stress fields are clearly visible around the θ-particles.

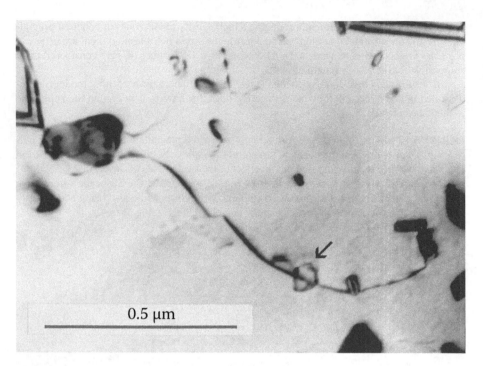

FIGURE 5.17 The cutting of coherent precipitates by dislocations is difficult to study because the sample is three-dimensional, while the recorded TEM image is a two-dimensional representation. Thus, superposition of features in the image can occur.

Low-carbon steel alloyed with vanadium. At first glance, the dislocation line is seemingly cutting through the precipitate indicated by the arrow. More probably it is sliding in a plane over or under the particle. Such a question can often be answered by taking stereo images of the region in question.

5.3   DISCONTINUOUS (CELLULAR) PRECIPITATION

Discontinuous precipitation, also called cellular precipitation, is another mode of transformation, through which the supersaturated solid solution lowers its free Gibbs energy. The decomposition of the supersaturated solid solution α results in transformed regions in the form of *cells (colonies, nodules)*, composed of lamellar precipitates of new equilibrium phase β and *depleted phase* α_l. The transformation reaction can be written as $\alpha \rightarrow \alpha_1 + \beta$. The phase α_1 has a lower thermodynamic excess of solute and the same crystal structure as the α-phase, but it has a crystallographic orientation different from the orientation of α in the grain, where the cell grows—that is, the α_1 phase inside the cell is reoriented. The two-phase cell (colony) is separated from the surrounding nontransformed parent phase α by a high-angle incoherent interface. The interphase boundary of the growing cell (the cell front) moves into the parent phase with the growing tips of the transformed phases and gradually consumes it, the growth being controlled by boundary diffusion through the colony/parent phase interface. The area of nonconsumed parent phase diminishes with time, thus preserving the initial supersaturation till the completion of the transformation.

The term *discontinuous* originates from the existing concentration gradient at the cell interphase boundary and the abrupt change of both composition (from α to α_1) and orientation of the parent phase when the advancing cell front passes through it. The term *cellular* reflects the morphological characteristics of the reaction products growing as a colony or duplex product regions (cell) with a constant interlamellar spacing, specific for the given temperature, between the two phases inside the colony.

Discontinuous precipitation has been observed in commercial Cu-Be, Cu-Ti, Mg-Al-Zn, Zn-Al, Pb-Sn alloys; nitrogen austenitic steels; and some magnetic alloys under certain, fortunately narrow, ranges of treatment conditions. The discontinuous precipitation reaction is usually unwanted during industrial aging treatments because the nucleation of the noncoherent phase on grain boundaries in the two-phase structure deteriorates the mechanical properties and causes brittleness and higher susceptibility to intercrystalline corrosion.

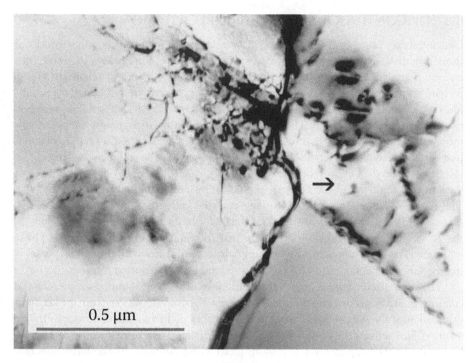

FIGURE 5.18

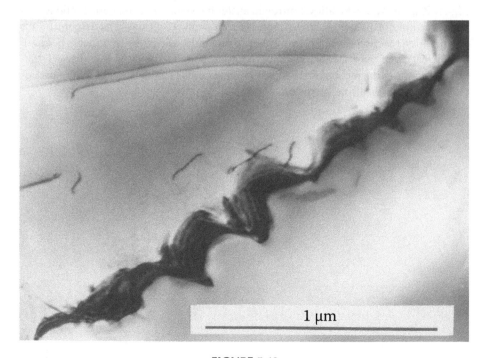

FIGURE 5.19

FIGURE 5.18   The discontinuous precipitation begins with nucleation of platelets of the new equilibrium phase on the boundary. The platelets lie facedown at the boundary; the other broad face of the platelet is semicoherent with the grain into which the precipitation will proceed.

Austenitic nitrogen steel Fe-18Cr-14Mn-0,6N quenched from 1100°C and aged at 650°C for 2 min. The cellular precipitation is observed in the high-nitrogen-alloyed stainless steels in the temperature range 550°C to 850°C. The first nuclei of the new phase (Cr_2N) precipitate on austenite grain boundaries within the first few minutes of aging. The growth direction is shown by the arrow into the austenitic grain.

FIGURE 5.19   The nuclei of the new phase precipitate on the grain boundary at roughly regular intervals and quickly cover its full length. In this way, the preferred nucleation sites are exhausted during the initial steps of the process.

Austenitic nitrogen steel Fe-18Cr-14Mn-0,6N quenched from 1100°C and aged for 5 min at 650°C. The austenitic grain boundary is almost completely covered by platelets of chromium nitride.

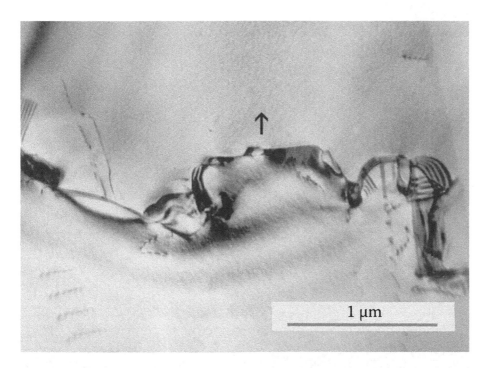

1 μm

FIGURE 5.20   The broad face of the nuclei of the plate-like phase is parallel to its habit plane. This forces the boundary to bow out between nuclei to remain parallel to this plane during the decomposition.

Austenitic nitrogen steel Fe-18Cr-14Mn-0,6N quenched from 1100°C and aged for 5 min at 650°C. The growth direction is shown by the arrow.

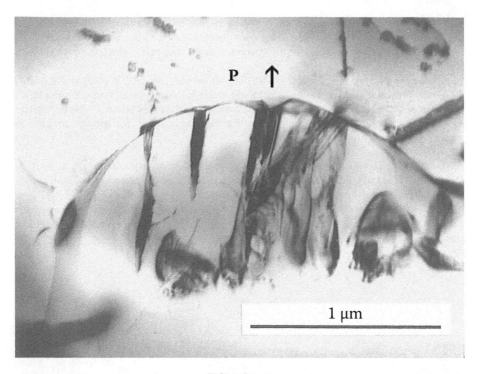

P ↑

1 μm

FIGURE 5.21

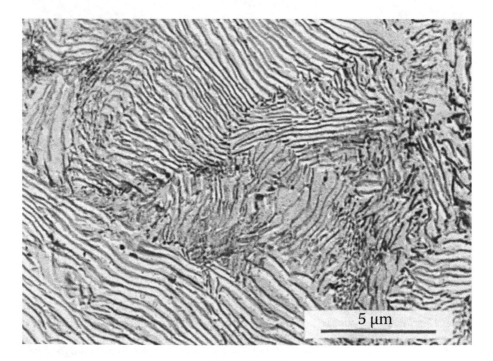

5 μm

FIGURE 5.22

FIGURE 5.21   Cellular decomposition continues through the cooperative growth of the new precipitated phase and the depleted phase, which are in the shape of parallel lamellae and form cells. The advancing cell front has the structure of a high-angle boundary, which provides effective diffusion path and solute transport even at relatively low temperatures. Thus, the growth rate of the phases is controlled by grain-boundary diffusion.

Austenitic nitrogen steel Fe-18Cr-14Mn-0,6N quenched from 1100°C and aged for 10 min at 650°C. The high-angle boundary between the cell growth front and the parent austenite phase is clearly visible through the high-angle crystallographic misorientation between the transformed austenite lamellae and the grain into which the colony grows, as marked by *P.*

FIGURE 5.22   The morphological characteristics of the discontinuously decomposed alloys are similar to those of colonies of pearlite.

Austenitic nitrogen steel Fe-18Cr-14Mn-0,6N quenched from 1100°C and aged for 60 min at 650°C. Replica. Compare this structure to the pearlite structure in Figure 5.27.

5.4   EUTECTOID TRANSFORMATION

Eutectoid reactions are reversible diffusional transformations that are induced by temperature transitions—cooling or heating through some transformation temperature, which is specific for each alloy system. On slow cooling below the eutectoid transformation temperature T_E, the metastable solid solution *decomposes* into a more stable mixture of two new solid *eutectoid phases*. The formation of phases of different composition is accompanied by a significant redistribution of the constituents, a process that requires *long-range diffusion*.

The eutectoid decomposition of the eutectoid alloy in a standard eutectoid-type phase diagram can be expressed in the form $\gamma_E \rightarrow \alpha + \beta$. The eutectoid morphology is often called *pearlite*, emphasizing with this terminology the main feature of the eutectoid structure—a configuration of alternate lamellae of two product phases.

Eutectoid transformations are found in many alloy systems: the ferrous systems Fe-C, Fe-N, Fe-Cu, and Fe-Nb and the nonferrous Cu-Sn, Cu-Al, Cu-Be, Cu-Si, Zn-Al, Ti-Cr, Ti-Fe, and Ti-Mn. From an engineering point of view, the eutectoid transformations in Fe-C alloys are the most important.

The nucleation of pearlite starts with the formation of a nucleus of either of the two product phases on boundaries of the parent phase (austenite in Fe-C alloys). Either of the two new phases can nucleate first, obeying a definite crystallographic relationship with one of the austenite grains. The nucleation of the second phase occurs adjacent to the first, producing an orientation relationship with it and with the same grain. Both nuclei have an incoherent mobile boundary with the other grain. The nucleation process is repeated sideways while the lamellae of the two phases grow cooperatively and form a *colony* (called also eutectoid grain) of alternating lamellae under this fixed crystallographic relationship. The colony grows radially into the parent austenitic grain with which it does not have an orientation relationship by a boundary diffusion mechanism through the colony/parent phase interface.

The morphologies of the microstructure resulting from the eutectoid decomposition and from the cellular precipitation are microscopically similar—two phases nucleate on grain boundaries and subsequently grow as a common aggregate of alternating lamellae. The main difference is that in the eutectoid reaction the parent phase transforms into a more stable mixture of two different phases, whereas in the cellular precipitation one of the phases in the cell is new, and the other is the same phase but reoriented and depleted.

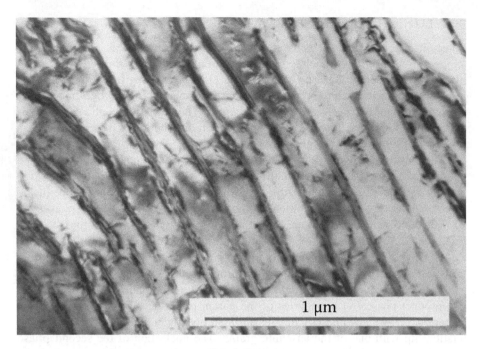

FIGURE 5.23

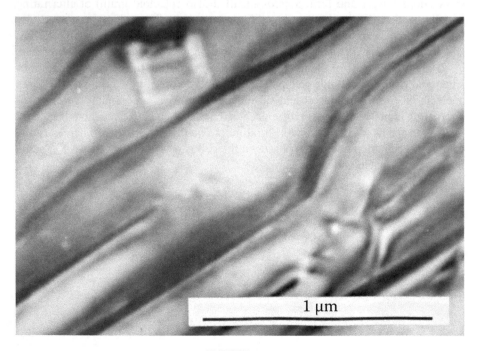

FIGURE 5.24

FIGURE 5.23 During the eutectoid decomposition, the parent phase (austenite) in ferrous alloys undergoes polymorphic transformation: the fcc austenite γ transforms into bcc ferrite α. The eutectoid decomposition in plain carbon steels proceeds according to the reaction $\gamma_{(0.8\%C)} \to \alpha + Fe_3C$ (cementite). The two-phase pearlite is composed of colonies comprising alternate lamellae of cementite and ferrite.

Pearlite in hypo-eutectoid carbon steel 0.25%C. The dark lamellae are cementite, and the light lamellae containing dislocations are ferrite.

FIGURE 5.24 Under isothermal conditions, the growing eutectoid lamellae in all colonies keep a roughly constant interlamellar spacing, which is specific for the transformation temperature and varies inversely with undercooling ΔT below the eutectoid transformation temperature.

Ferrite-pearlitic steel, containing 0.1%C, 0.1%N, and 0.1%V. The growing cementite lamellae change their direction when meeting primary vanadium carbonitride. The constant interlamellar spacing in the colony is restored by formation of additional cementite lamella.

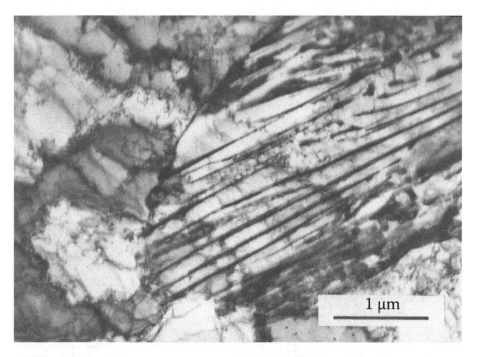

FIGURE 5.25 The rate of growth of the pearlite depends on the undercooling degree. In plain carbon eutectoid steels, the growth rate varies from 0.1 μm·s⁻¹ at lower ΔT to 100 μm·s⁻¹ at higher ΔT. The higher ΔT contributes to the better mechanical properties of pearlitic steels. This is especially true for the impact strength, which improves from both the smaller colony size and the smaller interlamellar spacing.

Cast carbon steel, containing 0.25%C. The relatively high transformation velocity resulted in eutectoid grains with diameters of several micrometers and a 50 to 100 nm interlamellar spacing.

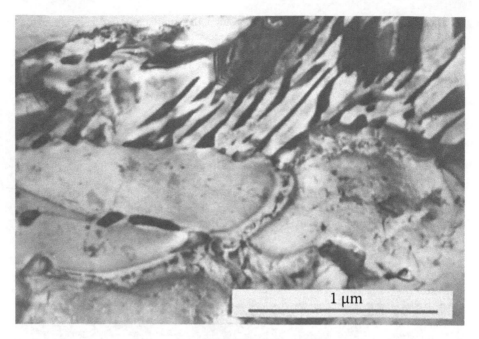

FIGURE 5.26

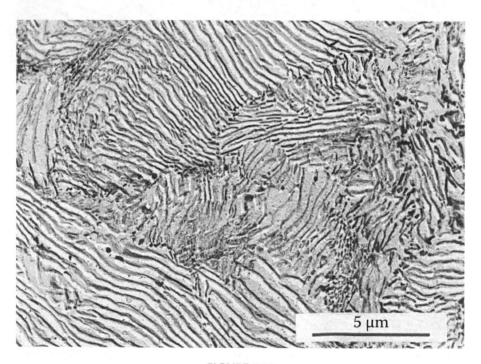

FIGURE 5.27

FIGURE 5.26 The amount of pearlite depends on the transformation time and on the carbon content of the steel. In hypoeutectoid steels, a precipitation of proeutectoid ferrite from austenite precedes the pearlitic decomposition of the remaining austenite, which has reached a eutectoid carbon content.

Low-carbon ferrite-pearlite steel, containing 0.1%C, 0.1%N, and 0.1%V. The ferrite grains contain a small number of short dislocation lines and fine carbonitrides. The high-angle ferrite/ferrite boundaries appear as two parallel lines where they intersect with the foil surfaces. Large cementite particles can be seen precipitated on these boundaries.

FIGURE 5.27 Another factor that influences the pearlite morphology and the pearlitic transformation kinetics, apart from the undercooling ΔT degree and chemical composition, is the size of the prior austenite grains. The smaller grains provide a bigger total length of grain boundaries, a higher number of nucleation sites, a smaller colony size, and a higher growth rate.

Plain carbon eutectoid steel. Replica. Pearlitic colonies have nucleated on all the austenite boundaries and have consumed the whole volume occupied by the fine grains of the parent austenite, thus producing 100% pearlite.

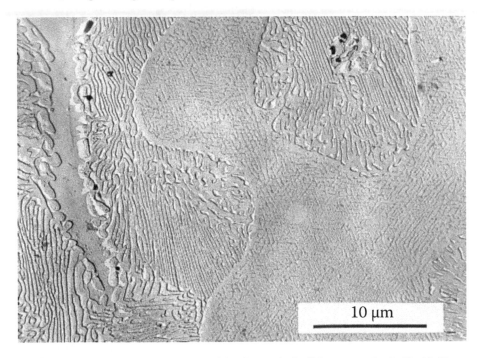

FIGURE 5.28 The eutectoid decomposition is a typical solid-state reaction in Zn-Al alloys. The pearlitic reaction in this case is $\beta \rightarrow \alpha + \eta$, where the high-temperature β-phase decomposes into α-solid solution of Zn in Al and a η-solid solution of Al in Zn.

Cast alloy Zn-27%Al. Replica. The eutectoid decomposition has produced colonies containing lamellae alternately of the α- and η-phases. The sites for the nucleation of the colonies are the grain boundaries and the interfaces of the large primary η-phase boundary precipitates. The rest of the volume has transformed into supersaturated α-solid solution, which has subsequently decomposed during cooling into fine η-needles. The Widmanstätten morphology of the secondary precipitated η-needles is due to the preferred nucleation on the $\{111\}_{Al}$ habit plane.

5.5 MARTENSITIC TRANSFORMATIONS

Martensitic transformations, frequently also called *shear* or *displacive transformations* (because their mechanism involves homogeneous lattice deformations and shear displacements), belong to the diffusionless solid-state transformations. No long-range diffusion is required for the nucleation and growth of martensite, and it inherits the chemical composition of the parent phase. The process does not rely on the assistance of thermal activation; still, for most of the alloys that transform martensitically, the speed of growth of the individual martensite plate is extremely great—it approaches the velocity of sound in solids. The diffusionless, athermal character of the transformation means that from start to completion the individual atoms move only a distance that is smaller than one interatomic spacing in the parent phase. These movements are cooperative and coordinated in a "military" manner that allows the atoms to retain their own original neighbors in the new phase. The military mode of movement of the atoms determines the crystallographic characteristics of the martensite transformation, which are the most distinguishable features from any other transformation. The coordinated structural change requires a fully coherent or semicoherent interface and leads to the establishment of an orientation relationship between the parent and the product lattices.

The term *martensite* was initially introduced to describe the hard structure obtained by rapid quenching of steel from the austenitic region. Nowadays, however, it is used in physical metallurgy as a generic name for the products of the diffusionless transformations. Martensitic transformations are observed in many pure metals with allotropic modifications (Co, Fe, Ti, Zr) and many important engineering materials such as carbon and alloyed steels; iron-based alloys (Fe-Ni, Fe-Mn, Fe-Cr); and nonferrous alloys, such as Cu-Al, $(\alpha+\beta)$-Ti alloys, and shape-memory effect alloys (Cu-Zn, Cu-Zn-Al, Cu-Zn-Ni, Ti-Ni). Martensitic transformations can also occur in some types of nonmetallic crystals, some minerals, ceramic inorganic compounds, and even polymers.

The best-known and undoubtedly most significant for engineering practice is the austenite-martensite transformation in Fe-C and Fe-Ni systems, which are basic for steels. This transformation occurs during rapid cooling (quenching) from the austenitic region to temperatures between M_s (the martensite start temperature) and M_f (the martensite finish temperature). Both M_s and M_f are definite temperatures for a given alloy composition but do not depend on cooling rate in a broad range of rates.

The martensite crystals nucleate in austenite as bcc, bct (body-centered tetragonal lattice), or hcp flat crystals. The product of the martensitic transformation in plain carbon steels is a metastable supersaturated solid solution of carbon in α-Fe having bct crystal lattice, the unit cell parameters and the degree of tetragonality (c/a ratio) depending on the carbon content.

The preferred habit plane, usually $\{111\}_\gamma$, and the strict crystallographic relationship between the martensite and austenite crystal lattices determine the planar semicoherent martensite/austenite interfaces. The morphological characteristics of bct and bcc martensite in quenched carbon steels and in carbon-free iron alloys have been adopted as a basis for the classification of martensite into two major types: *plate (twinned) martensite* and *lath (dislocation, massive, packet) martensite*. Depending on alloy composition, the two types differ in their shape, orientation variants, micro- and substructure, and habit plane.

The martensitic structures in nonferrous alloys are classified as plate martensite because of their plate-like shape. They are, however, further divided into two types based on their internal microstructure: *faulted* and *twinned martensite*. The twinned martensite is observed in many nonferrous alloys, such as Cu-Al (containing more than 9.5% Al), Cu-25%Sn, Au-Cd, and Au-Mn, and in Ti-based alloys with Cu, Fe, Mn, Mo, and V. Faulted martensite is observed in Cu-(11–13%)Al, Cu-(22–24%)Sn, Cu-(38–41%)Zn and in Co-based alloys with Fe, Ni, and Al. The *reversible thermo-elastic martensite* transformation running in Cu-Zn, Cu-Zn-Al, Cu-Zn-Ni, Ti-Ni, and Ti-Mo alloys is the physical ground of the *shape-memory* effect and the development of *"smart" alloys*.

FIGURE 5.29 *Plate martensite* morphology predominates in high-carbon steels, Fe-Ni alloys (containing more than 28% Ni), and some special steels. The plate martensite has mainly a lens or plate shape. The central part of the lens, called the *midrib* (abbreviation of *middle ribbon*), consists of fine parallel internal twins, which give the name *twinned* to this type of martensite. It has been suggested that due to the crystallographic match between the twins and the $\{111\}_\gamma$ habit plane, the midrib is the first part of the plate to grow. Toward the periphery of the plate, the twins degenerate into complex dislocation arrays.

Plate martensite in Fe-30%Ni alloy, quenched to -196°C (liquid nitrogen temperature). Note the zigzag array of martensite lenses, embedded in retained austenite. The direction of each midrib corresponds to the projection of a $\{111\}_\gamma$ habit plane.

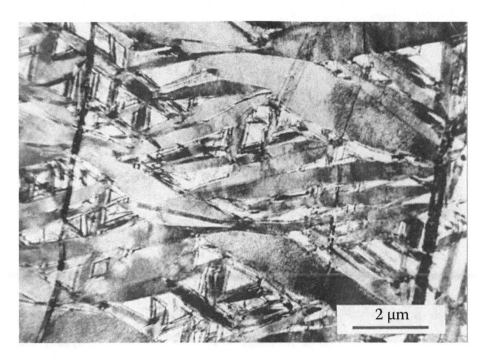

FIGURE 5.30   The typical microstructure after quenching to plate martensite consists of *martensite units*, composed of many zigzag-oriented plates. The length of the plates is restricted either by another martensite crystal or by an austenite grain boundary.

Units of plate martensite in Fe-30%Ni alloy, quenched to -196°C. Each grain contains several martensite units of different direction, but in all units the individual plates lie along the direction of the $\{111\}_\gamma$ traces.

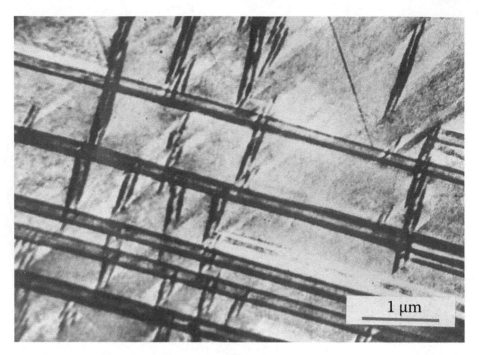

FIGURE 5.31

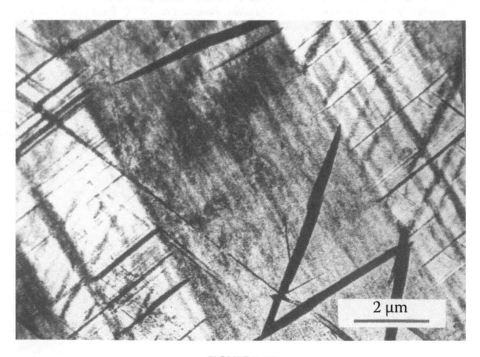

FIGURE 5.32

FIGURE 5.31   Plate martensite develops in some special high-carbon ferrous alloys during cooling to room temperature.

Cast Fe-3%C-30%Cr-15%Mn alloy. The plates of martensite are so thin that they could also be described as needles.

FIGURE 5.32   Needle-shaped plate martensite can be produced by deformation at high strain rates in many alloyed steels of lower carbon content.

Needle-like plate martensite in a shock loaded Fe-0.25%C-14%Mn-3%Cr steel. The martensite needles are surrounded by highly deformed austenite.

FIGURE 5.33   The cold plastic deformation of iron alloys of relatively high SFE can induce a transformation of the austenite into a *strain-induced martensite.*

Low-carbon steel with 0.25%C cold rolled to a 5% reduction at strain rate $\dot{\varepsilon} = 5\ s^{-1}$. The length of martensite needles is less than 1 µm, whereas their width is less than 10 nm. The interface strain fields and the dislocations accumulated around the needles result from the high strain rate.

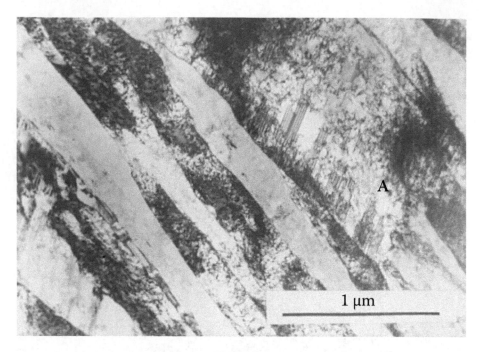

FIGURE 5.34

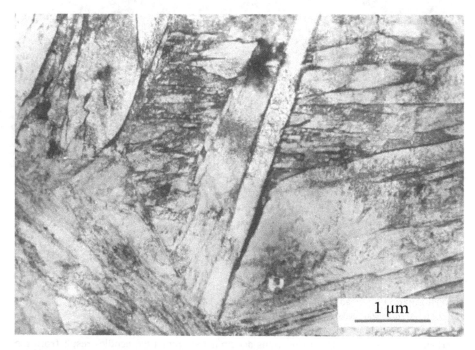

FIGURE 5.35

FIGURE 5.34   *Lath martensite* is the most common martensite morphology. It is observed in quenched low- and medium-carbon steels and in maraging and stainless steels. The lath martensite is referred to as *dislocation martensite* because of the dislocation substructure observed in the laths. It is similar to the substructure of cold-worked steels: the density of dislocations can be about 10^{11} to $10^{12}\,cm^{-2}$, and the dislocations form tangles that separate cells. Some of the tangles develop into low-angle or high-angle boundaries that border the elongated martensite laths. The boundaries are almost straight due to the preferred habit plane and the strict crystallographic relationship between the ferrite and the parent austenite.

Lath martensite in Fe-14Cr-5Ni-3Mo steel. The strong white-black contrast is an indication of the large misorientation angle across the lath boundaries. A roughly 1-µm-thick layer of retained austenite (marked by A) is easily distinguished by the bands of SFs in it.

FIGURE 5.35   The parallel martensite laths are grouped into *martensite units,* the units themselves into larger packets. The terms *packet* or *massive martensite* have been adopted for lath martensite, thus emphasizing the main morphological feature observed when studied in a light microscope.

Packet martensite in a plain carbon steel.

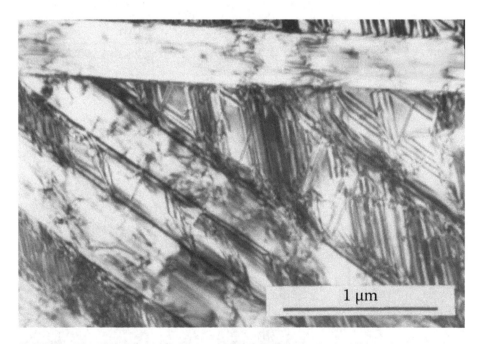

FIGURE 5.36   Cooling to very low temperatures and the plastic deformation of alloys of low SFE such as stainless steels, high-manganese Fe-Mn steels, and some complex alloyed steels induces the transformation reaction γ (fcc) $\rightarrow \varepsilon$ (hcp). The nonmagnetic ε-martensite is also called *sheet martensite* because it forms very thin sheets on $\{111\}_\gamma$ planes.

High-carbon Fe-30Cr-15Mn steel, quenched to room temperature. When observed in the TEM, the ε-martensite sheets produce the same contrast as SFs—alternating black and white and parallel fringes. The crystallography and the morphology of the SFs and of the ε-martensite are identical. They can be distinguished only by the greater thickness of the martensite sheets.

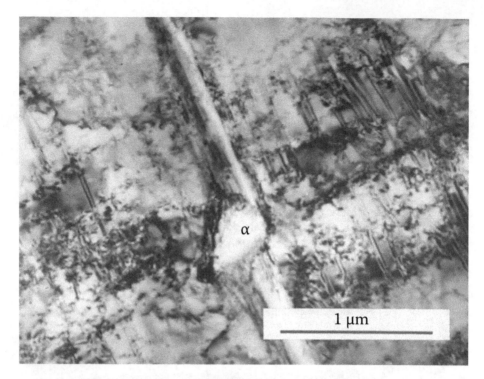

FIGURE 5.37

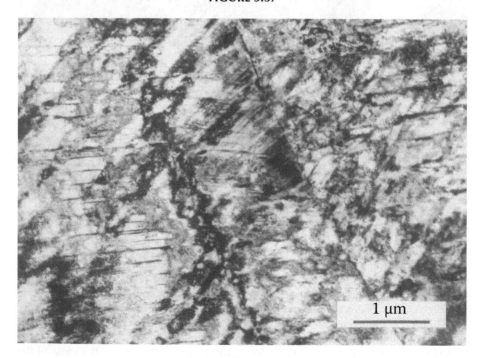

FIGURE 5.38

FIGURE 5.37 The arrangement of atoms produced by the crossing under certain crystallographic conditions of two sheets of ε-martensite corresponds to the arrangement of the bcc crystal lattice. These intersections are nuclei for the formation of bcc ferromagnetic α-martensite according to the reaction $\gamma \rightarrow \varepsilon \rightarrow \alpha$.

Austenitic Fe-18Cr-9Ni steel, 20% cold rolling. An α-martensite nuclei is indicated by the symbol α.

FIGURE 5.38 The amount of strain-induced ε- and α-martensite increases with the amount of deformation.

Fe-18Cr-9Ni stainless steel, 70% cold rolling. The fine martensite laths are embedded in highly faulted austenite.

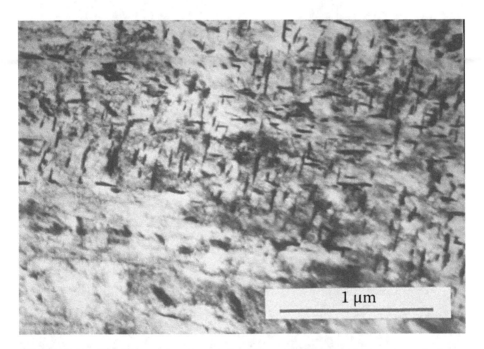

FIGURE 5.39 For most engineering steels and iron alloys, quenching to martensite is not a final heat treatment operation, and a *tempering* is applied. This is a heat treatment process of heating the martensitic steel to temperatures below Ac_1, aiming to improve toughness, ductility, and even strength. Various structural changes and solid-state reactions take place during tempering of ferrous martensite depending on the temperature and duration of the heat treatment, and on the steel composition. The most important are the segregation of carbon atoms to dislocations, the pre-precipitation clustering and the precipitation of carbides, the decomposition of retained austenite, the recovery of dislocation substructure, and the recrystallization of ferrite plates into recrystallized grains. Many of them overlap, leading to an extremely complicated microstructure.

Tempered martensite in a carbon steel. The fine needle-like carbides that have precipitated in the martensite plates are mutually oriented at an angle of 60°—this angle corresponds to that between $\{011\}_\gamma$ planes, which are the habit planes for cementite precipitation in martensite.

FIGURE 5.40

FIGURE 5.41

FIGURE 5.40   The plate martensite in nonferrous metals and alloys forms two morphologies that differ with respect to their internal structure, a feature that can be distinguished in the TEM. The twinned martensite forms plates that consist of fine twin bands separated by bands of nontransformed highly strained matrix.

Twinned martensite in a Cu-3Si-1.5Mn alloy (silicon bronze, Everdur) that was quenched from 900°C. The parallel twin and matrix bands are about 10 nm wide. Fine striations—SFs or secondary twin bands—can be seen in some of the matrix bands.

FIGURE 5.41   The fine structure of faulted martensite is usually defined as consisting of numerous internal faults, but the mechanism of their formation is still uncertain.

Faulted martensite in quenched Cu-26Zn-4Al alloy. The closely spaced striations running across the transformed grains are indicative of the presence of martensite with numerous internal faults, but their character is hard to determine unambiguously.

5.6   THE BAINITE TRANSFORMATION

The bainite transformation occurs in carbon steels at temperatures lower than that of the pearlite reaction and higher than that of the martensitic transformation. The reaction product, termed *bainite,* is a nonlamellar mixture of ferrite of low carbon supersaturation and carbides. The different morphology of the carbide distribution in the upper and the lower part of the bainite temperature range in steels has resulted in the classification of bainite into two groups: *upper bainite*, which is formed in the high-temperature part of the bainite range, and *lower bainite*, which is formed in the low-temperature part of the bainite range. There are differences in mechanical properties and microstructure of upper and lower bainite. Both upper and lower bainite form as aggregates of small plates or laths of ferrite (also called subunits). The carbide phase (cementite) in upper bainite precipitates from the carbon-enriched austenite predominantly between the plates of ferrite, the amount and micromorphology of cementite depending on the carbon content of the alloy. In lower bainite, there are two kinds of much finer carbide precipitates: carbide particles, which precipitate inside the laths of supersaturated lower *bainitic ferrite,* and cementite, which precipitates from the carbon-enriched austenite that separates the plates of ferritic bainite

The bainite reaction has some characteristics that are typical of the martensitic transformation. For example, the dislocation density in the α-phase of bainite is rather high, although generally not as high as it is in martensite. The bainite transformation produces on the prepolished surface relief effects, that are similar to those resulting from a martensitic transformation, especially in the case of lower bainite and tempered martensite.

On the other hand, as with the austenite-to-pearlite transformation, an incubation period exists for the bainite reaction. A C-curve can be constructed for the bainite transformation on a T-T-T isothermal transformation diagram. Both pearlite and bainite in steels appear as a two-phase product of iron carbide and ferrite. However, the ferrite and carbide phases in bainite do not grow cooperatively like they do in pearlite. The main difference between pearlite and bainite lies in their crystallography. The cementite and ferrite in the pearlite have no specific orientation relationship to the austenitic grain in which they are growing, whereas in upper bainite they do.

The terms *bainite transformation* and *bainite*, which were initially introduced to describe the reaction and its product found in steels during an isothermal transformation, subsequently have been expanded over the years to refer to some nonferrous alloys: β-brass, copper-aluminum, and tin bronzes. The only common feature between bainites in steels and bainites in nonferrous alloys is that they all arise from transformations that are not exclusively civilian or military in character but have some characteristics of both types.

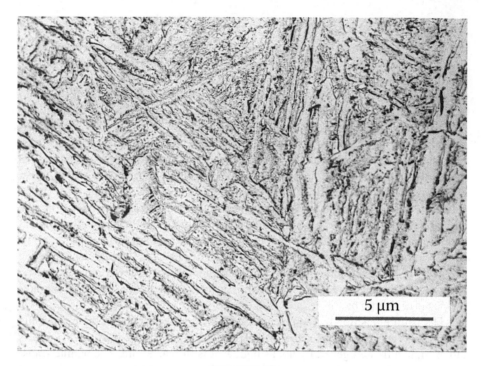

FIGURE 5.42

FIGURE 5.43

FIGURE 5.42   When observed at lower magnifications, the structure of lower bainite resembles the structure of lath martensite. The leading reaction is the transformation of austenite into supersaturated ferrite by a shear mechanism, which is typical of a martensitic transformation, followed by precipitation of very fine needle-like carbides in the interior of ferrite laths.

Lower bainite in carbon steel. Replica. Groups of parallel ferrite laths with preferred orientation along $\{111\}_\gamma$ form units in the parent austenite grain.

FIGURE 5.43   Higher-magnification studies reveal the main differences between lower bainite and martensite. The transformation twins that are typical for plate martensite are not present in bainite. The carbides in lower bainite are fine needles that precipitate inside the ferrite in parallel rows as in tempered martensite. In contrast to martensite, whose carbides precipitate on all close-packed crystal planes, lower bainite carbides precipitate predominantly in an identical orientation relationship to the ferrite, that is, about 60° to the direction of the ferrite plate.

Lower bainite in carbon steel. Compare this arrangement of carbides in lower bainite with the tempered martensite shown in Figure 5.39—the carbides in the martensite lath always show more than one orientation.

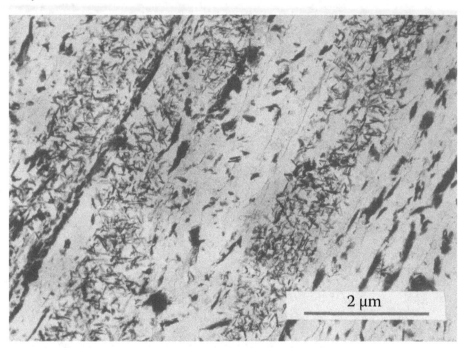

FIGURE 5.44   The differences between the distribution of carbides in lower bainite and martensite can clearly be seen in steels continuously cooled from the austenitizing temperature region at a rate intermediate between the martensite and bainite transformation rate and then tempered for a short time at temperatures close to the start of martensitic transformation.

Steel Fe-0.2C-2Cr, tempered for 8 min at 340°C. The structure is a mixture of plates of lower bainite and martensite. Rod-like carbides (cementite) with a size up to 0.5 μm are observed mainly on interlath boundaries of bainite and in smaller quantities inside the lath interior at an angle of 60° to the lath boundary. The numerous carbides in the adjacent martensite plate are fine needles up to 10 nm long, not more than several nanometers thick, and precipitated in three directions at 60° to the plate boundary.

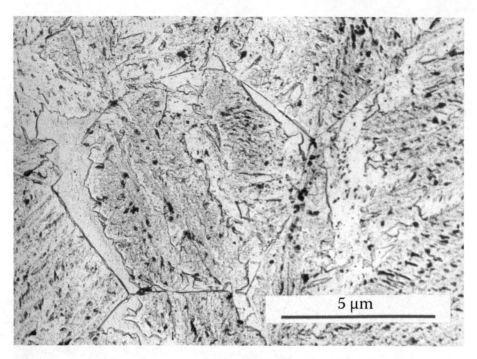

FIGURE 5.45

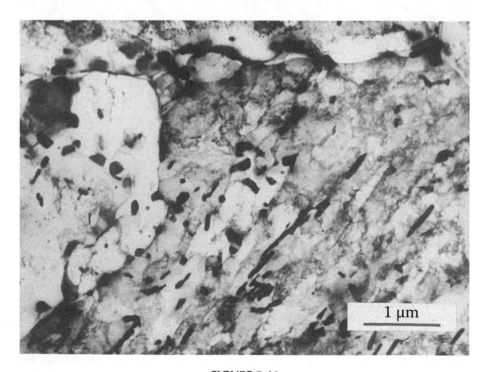

FIGURE 5.46

FIGURE 5.45   Upper bainite is an aggregate comprising ferrite laths and carbides precipitated mainly on lath boundaries. The parallel laths form plate- or feather-shaped regions in the grain. This structure is usually defined as *"feathery" bainite.*

Upper bainite in carbon steel. Replica. The grain boundaries of former austenite grains are covered by some retained austenite, these being the smooth areas in which no carbides are visible.

FIGURE 5.46   The density of dislocations in the interior of the upper bainite laths is rather high, but it is lower than in martensite. The type of boundaries separating laths is also different, being low-angle in the bainite and high-angle in the martensite. The most significant difference is the distribution of cementite particles (Fe_3C): in bainite, they are much larger, precipitate mainly in the interlath regions, and are less dense in the interior of the laths. In steels of higher carbon content, the particles can form nearly a complete carbide film between the laths.

Upper bainite in a carbon steel. High-angle boundaries of former austenite are visible in the upper part, whereas to the left of the micrograph they are covered by relatively large carbides.

6 Case Studies
Application of TEM in the Solving of Problems in Engineering Practice

Transmission electron microscopy has been applied for more than 50 years as a power-ful method for research purposes. Nowadays it is also recognized as a useful technique useful for the investigation of problems arising in engineering practice, especially for the development of new materials, the implementation of new methods for materials processing, and the identification of failures that occur during processing or service.

The main TEM applications are concerned with the following:

1. Phase transformations and structural changes occurring in metals and alloys as a result of the influence of temperature, rate of heating and cool-ing, and duration of the heat treatment
2. Structural changes occurring during the aging of alloys: precipitation kinetics plus the type, shape, size, volume fraction, and distribution of the strengthening phases
3. Deformation behavior and processes determined by the formation and sub-sequent changes of the dislocation structure depending on the deformation conditions
4. In-service behavior of alloys—changes in their microstructure under the influence of applied stress, temperature, or environment

6.1 DEFORMATION BEHAVIOR OF NICKEL-SILVER ALLOY IN THE TEMPERATURE RANGE FROM 100°C TO 900°C

A process of edge cracking has been observed during the processing of CuNi10Zn36Mn alloy (Nickel-Silver 10) by hot rolling. Metallographic, scanning electron microscopy (SEM), and TEM observations and mechanical properties test-ing at elevated temperatures were performed to find the causes for the edge cracking occurrence and to offer a solution to overcome the problem. It was found that a char-acteristic feature of the deformation behavior of the alloy in the temperature range from 100°C to 900°C was the sharp loss of ductility at 500°C to 550°C. The frac-ture in this temperature range had a mixed character—brittle and ductile. A TEM investigation of the dislocation structure and the precipitation processes as well as their associated changes over the temperature range from 100°C to 900°C helped to clarify the cause of the minimum in the ductility.

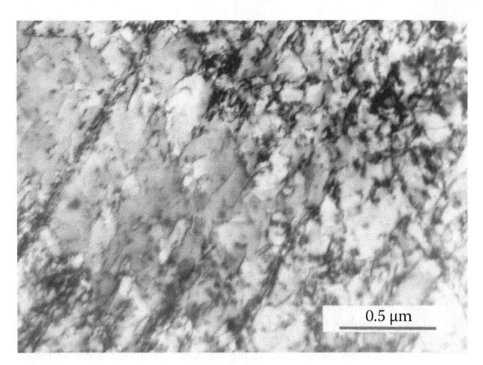

FIGURE 6.1

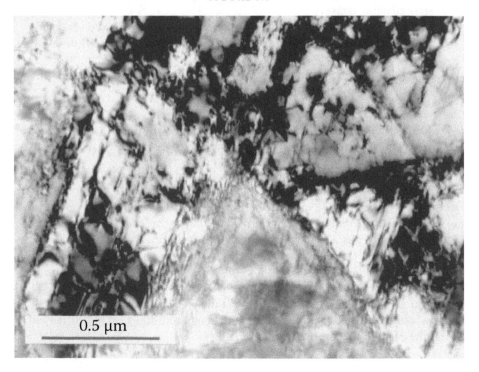

FIGURE 6.2

FIGURE 6.1   *The distribution of dislocations at 100°C is typical of metals of low SFE.* The partial dislocations which are separated by ribbons of SF glide in arrows on the close-packed crystal planes of copper.

FIGURE 6.2   *The planar distribution of dislocations is preserved up to 500°C.* The thermally assisted interaction and recombination of dislocations in this temperature range results in formation of deformation bands on the close-packed slip planes.

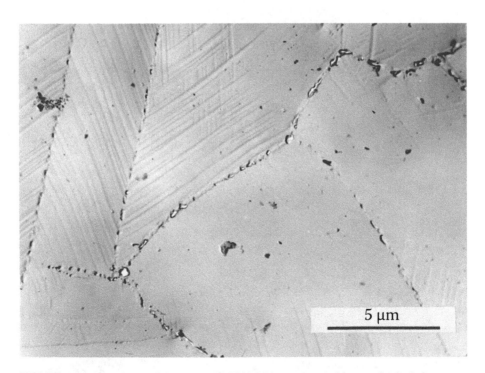

FIGURE 6.3   *The plastic deformation at 500°C is accompanied by accelerated phase precipitation.* Numerous particles of tetragonal intermetallic θ-phase (MnNi) precipitate and form chains on both high-angle grain boundaries and coherent and noncoherent twin boundaries, as can be seen in this replica. The precipitates impede the migration of the boundaries, thus lowering the alloy ductility at this temperature.

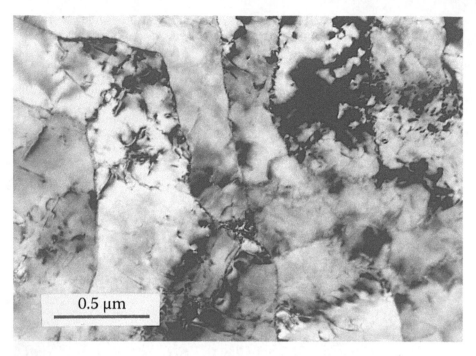

0.5 μm

FIGURE 6.4

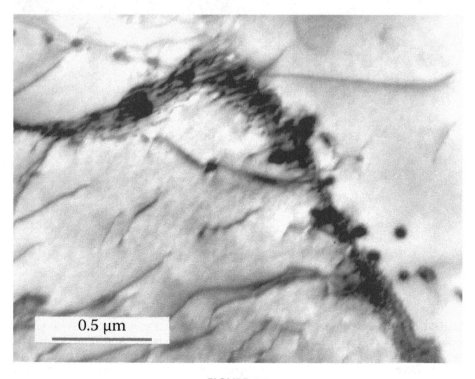

0.5 μm

FIGURE 6.5

FIGURE 6.4 *At temperatures of 600°C to 700°C, the SFE of the alloy increases and the dislocations move as unit ones.* The easy cross-slip and the interaction of the dislocations facilitated by the thermal activation result in the formation of subboundaries and a lower dislocation density. The enhanced diffusion in this temperature range provides for a high mobility of the boundaries and subboundaries, which then are capable of detaching from the precipitates. As a consequence, the ductility of the alloy is restored.

FIGURE 6.5 *Very active diffusion processes accompany the deformation at 800°C.* The structure is characteristic of dynamic recrystallization: grains of very low dislocation density separated by high-angle boundaries. Isolated θ-particles are present in some of the grains, but because they are not connected to grain boundaries, they cannot impede the migration of the boundaries.

CONCLUSIONS

It was proven that the reason for the loss of ductility and resulting edge cracking was the drop of temperature at the end of the rolling to the temperature range of low ductility. The detrimental effect of the precipitation-induced grain boundary embrittlement was eliminated and the hot workability of the alloy improved by a combination of a grain refinement (modification treatment) and optimization of the hot-rolling temperature-deformation procedure.

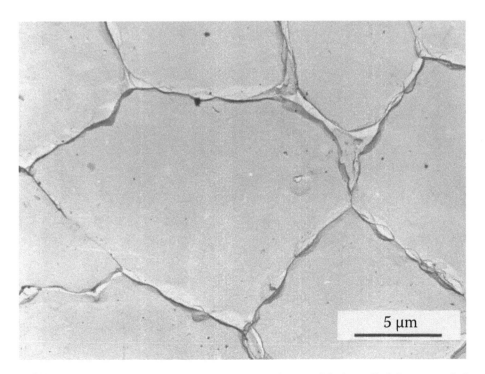

FIGURE 6.6 *Replicas prepared from the edge crack zone of the hot-rolled sheets revealed a continuous grain boundary film of coarse precipitates of θ-phase MnNi.* This structure corresponds to the structure produced in this alloy by deformation in the low-ductility region covering the temperature range from 500°C to 550°C.

6.2 DISTRIBUTION OF STRENGTHENING PHASES IN PRECIPITATION-HARDENING ALLOYS

The properties of heat-treatable precipitation-hardening alloys are strongly dependent on the distribution of the strengthening phases. The industrial interest in developing technologies for aging and thermomechanical treatments that produce uniform, finely dispersed precipitates in engineering alloys results from the requirements of achieving not only improved mechanical properties but also better corrosion behavior, toughness, fatigue resistance, and so on.

The *solidification conditions* are of great importance to the microstructure and properties of heat-treatable cast alloys. They determine both the type and the morphology of the primary intermetallic phases and thus affect the selection of proper processing parameters (solution treatment and aging temperatures and times) upon which the aging response and the type, distribution, and volume fraction of the strengthening phases depend. The optimal *heat treatment* regimes should be combined with the proper deformation conditions to provide the optimum microstructure and properties of wrought products for a given application. For example, a structure that contains a high density of uniformly distributed coherent or semicoherent fine precipitates provides high strength to aged alloys, whereas the absence of large precipitates on grain boundaries ensures good resistance to intercrystal corrosion, high ductility, and fracture toughness.

The appropriate *thermomechanical treatment* selected for a given alloy is another efficient way to obtain a finer uniform distribution of intermediate precipitates, because the increased dislocation density provides more sites for heterogeneous nucleation of these particles.

TEM studies of the structure of dispersion-hardened alloys facilitate the development and the introduction in practice of new optimal regimes for aging and thermomechanical treatment because they permit the evaluation of the influence of the different processing parameters on the microstructure.

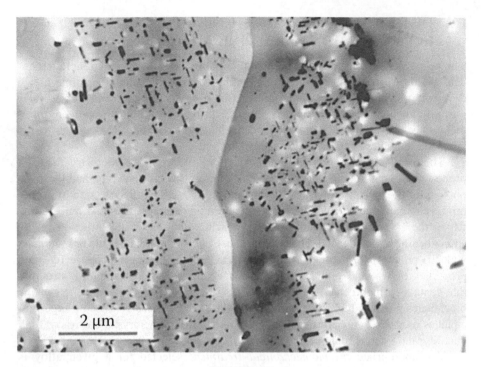

FIGURE 6.7

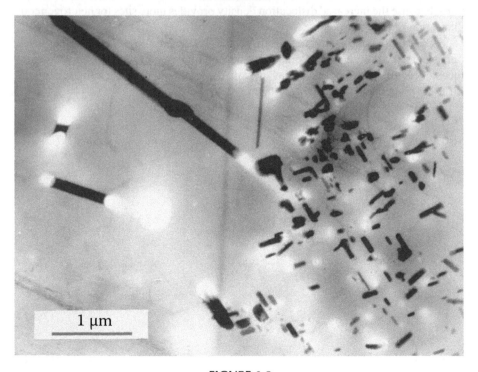

FIGURE 6.8

FIGURE 6.7 Frequently the aging treatment does not produce the desired uniform distribution of strengthening phases. For example, aluminum alloys suffer from the appearance of precipitation-free zones (PFZ)—regions adjacent to grain boundaries where no precipitation has occurred.

Alloy Al-6%Cu-0,2%Ti-0,2%Zr, mold casting with air-cooling, followed by 5 h solution-treatment at 540°C and 12 h aging at 160°C.

FIGURE 6.8 The air-cooling of cast aluminum alloys results in precipitation of coarse intermetallic phases on grain boundaries and solidification of coarse iron-containing needle-like compounds in the grain interior. The coarse needle-like phases are especially dangerous to the mechanical properties of all cast aluminum alloys because they act as stress concentrators and provoke nonuniform distribution of the secondary strengthening precipitations. Most often the inhomogeneous distribution of strengthening phases after aging is due to uneven supersaturation caused by coring and improper quenching regime.

Alloy Al-6%Cu-0,2%Ti-0,2%Zr, mold cast with air-cooling, followed by 5 h solution-treatment at 540°C and 12 h aging at 160°C.

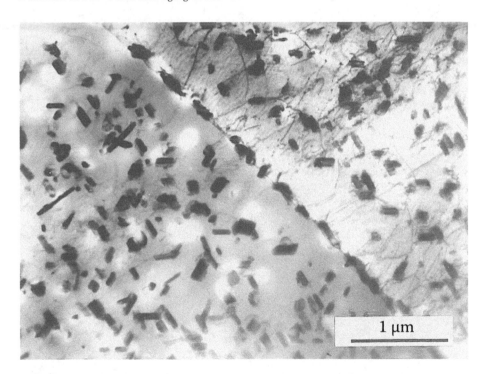

1 μm

FIGURE 6.9 The amount and the distribution of primary intermetallic phases can be managed by controlling the cooling rate through the mode of casting. Rapid cooling from temperatures just below the solidification temperature results in refined primary precipitates, a higher supersaturation, and a more homogeneous distribution of alloying elements retained in the solid solution. This provides for a shorter aging time and a better aging response, as well as avoiding the formation of unwanted PFZs.

Alloy Al-6%Cu-0,2%Ti-0,2%Zr, die cast and rapidly cooled from 540°C, followed by 5 h solution-treatment at 540°C and 12 h aging at 160°C. The uniform distribution and similar size of the secondary precipitates on the grain boundaries and in the grain interior ensure better mechanical properties compared to the air-cooled casting from the same alloy.

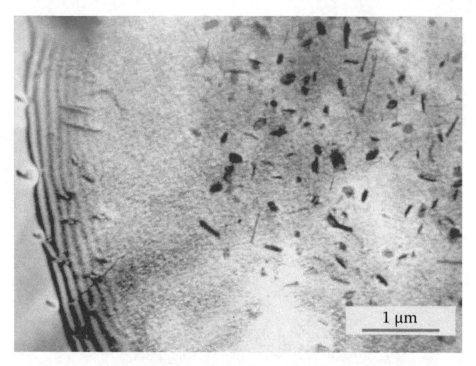

FIGURE 6.10

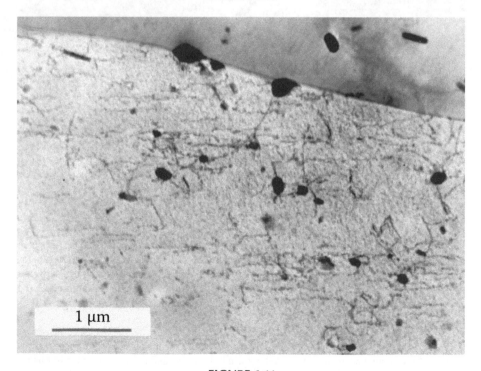

FIGURE 6.11

FIGURE 6.10 The high mechanical strength of this typically wrought aluminum alloy is provided by the disperse precipitation of a rodlike β′-phase (Mg_2Si) and the zone precipitation in the grain interior. The size of the primary intergranular Mg_2Si precipitates is relatively small, but a PFZ about 2 μm wide is seen around the grain boundary. The precipitation of zones occurs throughout the entire grain, including the depleted PFZ region, and this indicates that the zones are composed of mainly copper and chromium atoms.

Alloy 6061 (Al-1%Mg-0.6%Si-0.2%Cr-0.27%Cu), die casting followed by 2 h solution-treatment at 510°C and 12 h aging at 160°C.

FIGURE 6.11 The cold working of a quenched alloy provides numerous sites for subsequent heterogeneous nucleation of intermediate phases. This effect is used in thermomechanical treatments where a controlled degree of deformation is applied prior to aging to improve the properties as well as to accelerate the aging process and to minimize the possible deleterious effect of PFZs in the alloy.

Alloy 6061 (Al-1%Mg-0.6%Si-0.2%Cr-0.27%Cu) cold-forged and then aged for 12 h at 160°C. The heterogeneous precipitation of Mg_2Si in the deformed structure is more uniform, and there are no PFZs. The diminished size of zones, the change of the shape of rodlike Mg_2Si particles to a rounded form, and the presence of some dislocations in the aged structure all contribute to the improved properties of the thermomechanically treated alloy.

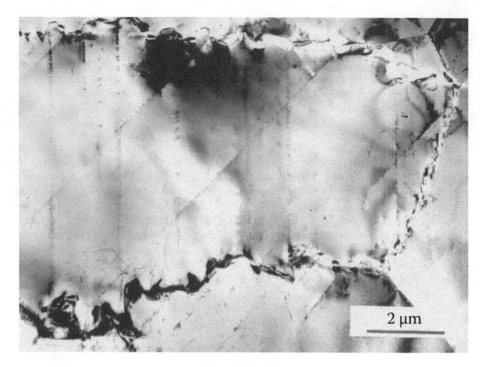

FIGURE 6.12

FIGURE 6.13

FIGURE 6.12   The preferential precipitation of phases on the grain boundaries greatly deteriorates the mechanical properties of alloys and can promote intergranular corrosion. This is a serious problem for the high-chromium austenitic stainless steels in the "sensitizing" temperature range from 500 to 900°C in that chromium-rich phases (carbides, nitrides, intermetallides) precipitate on grain boundaries.

Austenitic nitrogen steel Fe-18Cr-14Mn-0,6N, quenched from 1150°C and then tempered for 30 min at 700°C. The most dangerous temperature range for the high-nitrogen austenitic steels lies between 550°C and 850°C, when, supersaturated by nitrogen, austenite decomposes by a discontinuous mechanism. The process starts with the precipitation of Cr_2N particles on grain boundaries (see Chapter 5, Section 5.3).

FIGURE 6.13   The numerous defects of crystal structure introduced by plastic deformation prior to aging serve as centers for second-phase nucleation and can change the precipitation mechanism from localized grain-boundary precipitation to uniform continuous precipitation. Suppression of cellular precipitation by applying severe cold plastic deformation prior to aging is also observed in nonferrous alloys—for example, copper-beryllium.

Austenitic nitrogen steel Fe-18Cr-14Mn-0,6N, quenched from 1150°C, cold worked with 75% reduction, and tempered for 10 min at 700°C. The severe cold work stimulates the precipitation of uniformly distributed globular nitrides in the grain interior. The large number of particles provides numerous recrystallization centers and hence retards grain growth. The final structure of ultra-fine grains and uniformly distributed nitrides provides excellent mechanical properties and good corrosion resistance to the material.

6.3   SPECIFIC FEATURES OF THE STRUCTURE DEVELOPED DURING DEFORMATION IN SUPERPLASTIC STATE

The ability of metallic material to exhibit an extensive plastic deformation—up to 1000% elongation without work hardening—is termed *superplasticity.* Superplasticity is a state used for applying superplastic forming (SPF) and hot sizing to fabricate complex objects and parts for aerospace, automotive, and other applications.

Superplastic behavior is observed in metallic materials in two cases: if they possess an ultra-fine-grained structure (in the case of *structural superplasticity*) or if they undergo a phase transformation during deformation or thermocycling (in the case of *transformation superplasticity*).

The basic deformation mechanism of structural superplasticity is grain boundary sliding. Several preconditions are required to achieve this type of superplasticity: (a) the structure should be fine-grained with high-angle grain boundaries; (b) the grain size should vary from 0.5 μm to not more than several micrometers, a value comparable with the size of subgrains; and (c) the deformation should be carried out under special temperature-strain rate conditions.

The examination of the structural phenomena that underlie the plastic deformation in the superplastic state has to be undertaken on TEM thin foils due to the very finely dispersed structure.

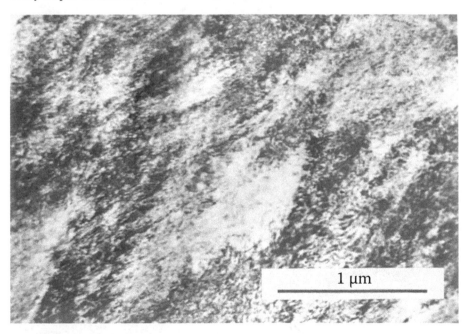

FIGURE 6.14   One of the most frequently used methods to produce an ultra-fine initial structure suitable for deformation in superplastic condition is severe cold deformation followed by a controlled recrystallization annealing. The severely deformed structure provides numerous nucleation sites which are necessary for the future homogeneous precipitation of a second phase.

Austenitic nitrogen steel Fe-18Cr-14Mn-0,6N, after cold rolling to a 75% reduction. The severe plastic deformation ensures the necessary extremely high dislocation density.

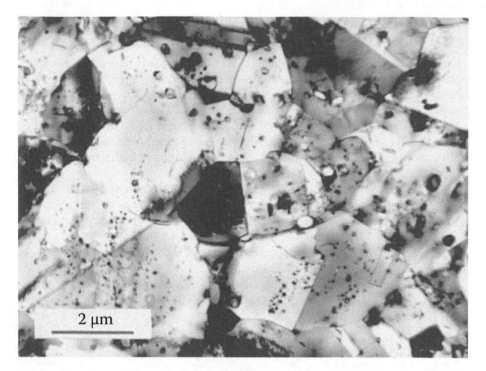

FIGURE 6.15

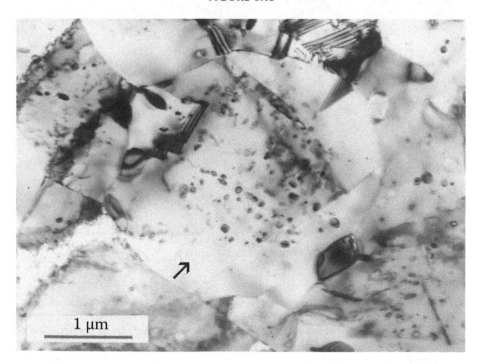

FIGURE 6.16

FIGURE 6.15 *Numerous nitride particles precipitate uniformly during the annealing for 60 min at 800°C of the severely deformed austenitic nitrogen steel Fe-18Cr-14Mn-0,6N.* The particles ensure numerous recrystallization centers and simultaneously act as barriers for grain growth. As a result, the average diameter of the recrystallized grains is about 3 to 4 μm. The obtained structure is suitable for further deformation in superplastic conditions.

FIGURE 6.16 *The fine-grained austenitic nitrogen steel Fe-18Cr-14Mn-0,6N was tensile loaded at a strain rate of $\dot{\varepsilon} = 1.10^{-3}\ s^{-1}$ at 800°C up to 100% elongation without fracture.* It can be seen from the micrograph that the Cr_2N particles preserve their size during the deformation, thus impeding the migration of grain boundaries and ensuring the stability of the grain size. Note the zones free of particles around grain boundaries that are normal to plastic flow direction (shown by arrow). They arise as a result of the accommodation processes that run in the regions around boundaries but not in the grain interior. Another unique feature of the deformation in the superplastic regime is the absence of dislocations. This shows that the main deformation mechanism does not rely on dislocations in the grain interior.

1 μm

FIGURE 6.17 *The elongation of the fine-grained austenitic nitrogen steel Fe-18Cr-14Mn-0,6N under deformation at the same conditions as in Figure 6.16 reached 400% before fracture.* The observation of thin foils taken from the region just below the fracture showed the presence of only a few dislocations in the interior of the grains. New elements visible in the micrograph are the slight grain growth and the appearance of vacancy complexes or loops in the interior of the grains.

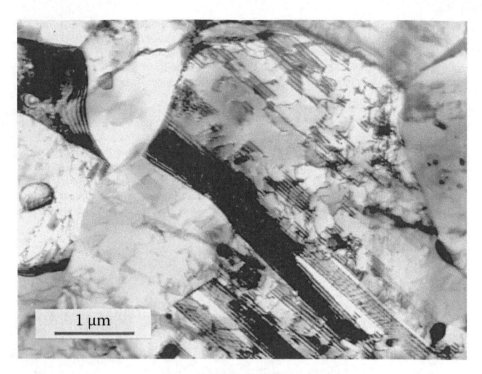

FIGURE 6.18

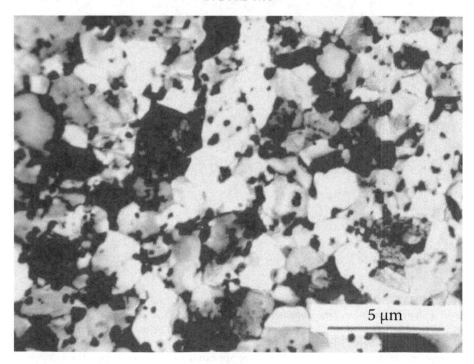

FIGURE 6.19

FIGURE 6.18 *The superplasticity is manifested only in a definite temperature-strain rate deformation range, which is specific to a given alloy.* Deformation beyond this range involves grain growth and grain elongation parallel to the flow direction, increase of dislocation density in the grains, and work hardening. These features, which are typical for conventional deformation at elevated temperatures, can be seen in the micrograph of the same fine-grained austenitic nitrogen steel Fe-18Cr-14Mn-0,6N, which was deformed at the same temperature of 800°C but at a higher strain rate of $\dot{\varepsilon} = 1.10^{-2}$ s^{-1}. The fracture has occurred at 80% elongation. The elongated grains with dislocations in their interior are a definite indication of deformation outside the optimum range for superplasticity.

FIGURE 6.19 *Another method to produce structure with ultra-fine grains necessary for the superplastic deformation is through the thermomechanical treatment of alloys.* This method proved itself suitable for the fine-grained Zn-1.2%Mn alloy, in which an average grain size of about 1 μm was obtained by extrusion to an 85% reduction at 275°C. The grain refinement was achieved through the precipitation of a ζ-phase (MnZn$_{13}$ intermetallide) during the deformation.

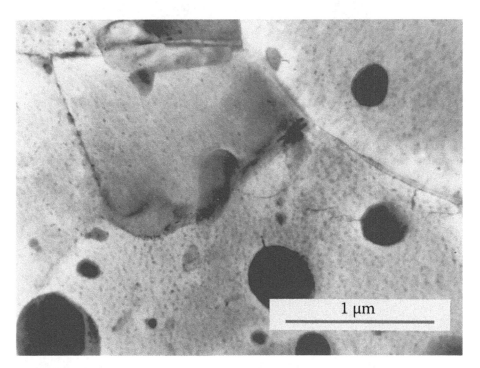

FIGURE 6.20 *The deformation of the ultra-fine-grained Zn-1,2%Mn alloy at a deformation rate $\dot{\varepsilon} = 1.10^{-3}$ s^{-1} at 300°C ensured an elongation of 680%.* The observations of thin foils showed that the principal phenomena controlling the superplastic deformation in these conditions was the plastic flow in the grain-boundary regions. As in the fine-grained austenitic steel (see Figures 6.9 and 6.10), the grain interior of the Zn-1.2%Mn alloy remains almost free of dislocations. The single dislocations present are usually bound to the interfaces of larger MnZn$_{13}$ particles. This shows that the flow is realized mainly by grain boundary slip and that the interfaces of larger precipitates take part in the accommodation process also.

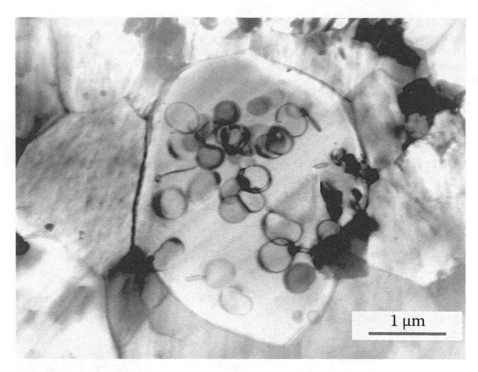

FIGURE 6.21 *A typical feature of the microstructure of ultra-fine-grained Zn-Mn alloys when deformed at temperatures lower than or at a strain rate higher than the optimum for superplasticity is the appearance of vacancy loops in the grains.* The appearance of point defects can be explained by insufficient diffusion flow that is unable to ensure the complete accommodation of adjacent grains. This is demonstrated in the micrograph of an ultra-fine grained Zn-0.7%Mn alloy after deformation at a deformation rate $\dot{\varepsilon} = 1.10^{-3}$ s^{-1} at 300°C. The maximum elongation was only 60%. Note that the shape of MnZn$_{13}$ particles is not rounded but somewhat faceted and that their size is much larger than in the Zn-1.2%Mn alloy (see Figure 6.19). Obviously the applied temperature-strain rate interval, which was optimal for the Zn-1.2%Mn alloy, is not suitable for the lower manganese alloy to reach superplastic state.

6.4 THE INFLUENCE OF MODIFICATION AND HEAT TREATMENT ON THE MICROSTRUCTURE OF ALUMINUM-SILICON ALLOY

Proper control over solidification is a first step toward controlling the metallurgical structure, metallurgical quality (soundness, homogeneity), mechanical strength, and physical properties of the castings.

Modification is a term used to denote the addition of "modifying agents"—small quantities of certain elements—to the molten metal to produce melts of better fluidity and feeding characteristics, to achieve better production rates, and finally to obtain castings with improved quality and mechanical properties.

The refinement of the as-cast grain size has proved to bring great benefits to numerous cast and wrought alloys. *Grain refinement* is a term that comprises the decrease in the size of grains or of dendritic cells by controlled additions to the molten metal before casting of "grain refiners," that is, selected elements in the form of master alloys.

The modification is widely applied to the most important commercial cast Al-Si alloys. In this case, the term *modification* refers to the process of treating or cooling the metal in such a way that the eutectic silicon grows as small, rounded fibrous particles instead of coarse, angular primary grains.

The effect of the cooling rate is limited within the thick sections of the castings, especially in sand or permanent molds. In this case, the modification is accomplished by elemental additions. Foundries have recognized for many years sodium metal or sodium salts and now strontium to be the best modifiers for eutectic and hypoeutectic Al-Si alloys. For example, modification is used to increase the strength and ductility of the non-heat-treatable AlSi12 alloy. Heat treatment of modified Al-Si-Mg alloys provides a further increase in strength.

The structure of Al-Si alloys is usually studied using TEM replicas. This is due to the large difference in the electrochemical properties of the Si and Al phases, which prevents thin foils of good quality from being obtained by the conventional electropolishing technique. Nowadays the invention of ion milling has overcome this limitation.

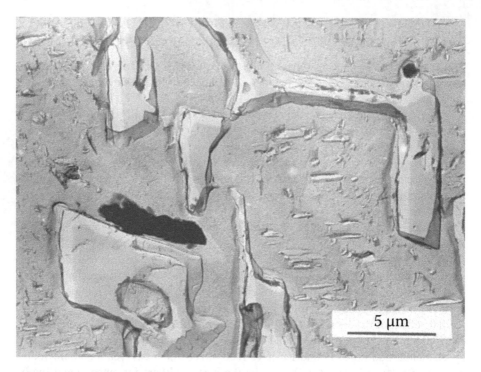

FIGURE 6.22

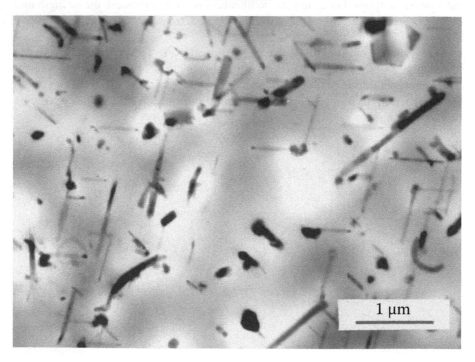

FIGURE 6.23

FIGURE 6.22 *The combined effect of modification and heat treatment on the microstruc-ture was studied in a heat-treatable A1-12%Si alloy containing additions of Mg, Cu, Ni, and Cr.* The alloy was developed for counter-pressure casting and intended for service at elevated temperatures. The slow cooling of all cast, nonmodified Al-Si alloys produces, as can be seen in the replica, large dendritic α-$_{Al}$ solid solution cells surrounded by coarse eutectic struc-ture built of plate-like silicon particles. The fine needles in the cell interior are intermetallic phases that have precipitated during the slow cooling of the casting.

FIGURE 6.23 *This micrograph shows a rare observation of small areas of the α_{Al}-solid solution in electropolished thin foil of the same alloy as in Figure 6.22.* Several dispersion phases act as strengthening agents: $CuAl_2$, Mg_2Si, Cu-Mg-Al, and Al-Cu-Ni. They all precip-itate from the supersaturated α-$_{Al}$ solid solution in needle-like form and can be differentiated in the TEM using electron diffraction, energy dispersion analysis (EDS), or electron energy loss analysis (EELS). The heat resistance of the alloy is provided mainly by the complex nickel-containing phases.

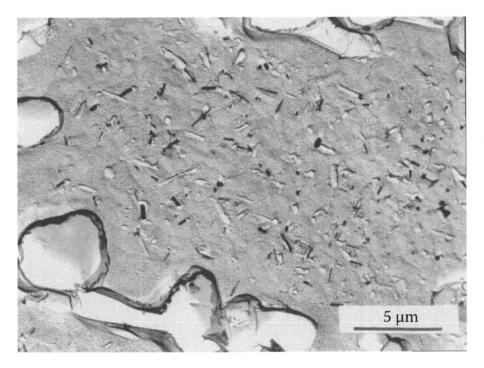

5 μm

FIGURE 6.24 High-temperature solid solution heat treatment prior to quenching of all nonmodified Al-12%Si alloys leads to some spheroidization of the silicon particles in the eutectic, but does not influence their size or the dendritic cells' diameter.

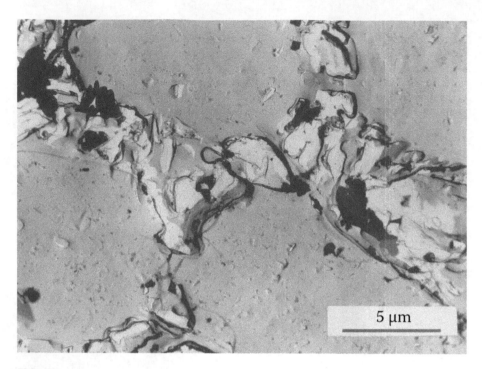

FIGURE 6.25 *The addition of specific quantities of modifying agents—in the form of an Al–Sr-master alloy or sodium salts—alters the shape of the coarse plate-like silicon particles to a rounded form while also restricting their growth.* In addition, it causes refinement of dendrite cells. The modified Al-Si eutectic and the refined dendritic cells result in an improved yield strength and especially in an improved elongation of the alloy.

6.5 SIGMA-PHASE FORMATION IN A DUPLEX STAINLESS CHROMIUM-MANGANESE- NITROGEN STEEL

The prolonged holding of high-alloyed stainless steels at elevated temperatures causes the precipitation of intermetallic compounds, which withdraw a significant amount of the alloying elements (Cr, Mo, Ni) from the austenite. The most serious result of this process is the depletion of austenite of chromium, the latter being the main element that provides the corrosion resistance of the steels. The precipitation of chromium-rich phases occurs at grain boundaries, thereby deteriorating both the mechanical properties and the resistance to intergranular corrosion. The intermetallic phases, which can form in the high-alloyed stainless steels, are designated with Greek letters: σ (sigma), ς (zeta), and χ (chi). A typical member of this group of phases is the σ-phase—a hard and brittle Fe-Cr intermetallide with a stoichiometry that depends on the chemical composition of the steel.

The mechanism of σ-phase formation is presented here for the duplex Fe-18%Cr-12%Mn-0.03%C-0.2%N steel, but it is representative for all high-chromium-containing stainless steels.

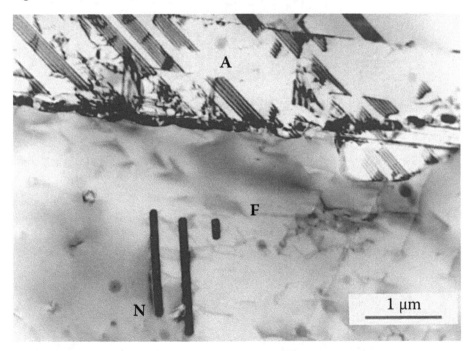

FIGURE 6.26 The structure of quenched nitrogen-alloyed Cr-Mn steels depends on the nitrogen content: they are austenitic at nitrogen content higher than 0.3% and ferrite-austenitic at lower concentrations. The equilibrium solubility of nitrogen in α-iron at 1100°C is less than 0.001%N, which means the ferrite of the low-nitrogen steels is supersaturated with nitrogen and easily decomposes on quenching.

Fe-18Cr-12Mn-0,03C-0,2N steel, quenched from 1150°C. Needle-like Cr$_2$N particles (marked by *N*) have precipitated from ferrite (marked by *F*) during quenching. Austenite (marked by *A*) has the typical structure of a low-SFE alloy—planar distribution of partial dislocations, separated by ribbons of SFs.

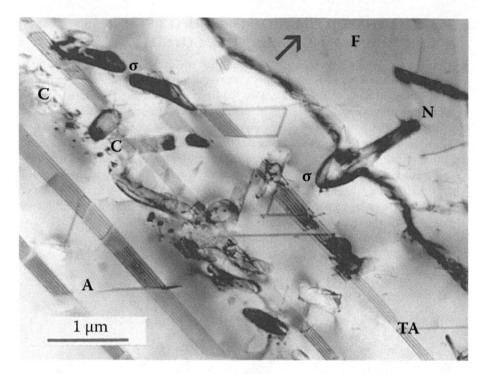

FIGURE 6.27

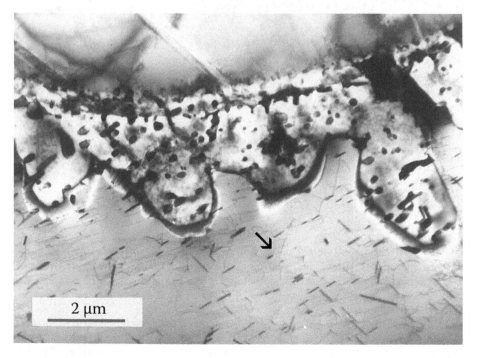

FIGURE 6.28

FIGURE 6.27   When the steel is heated and then held for a prolonged period at a temperature in the range from 550°C to 850°C, disperse chromium-rich $M_{23}C_6$ particles precipitate
on the ferrite–austenite boundaries. They consume the chromium from the boundary region,
thus inducing transformation of adjacent ferrite (F) into austenite (A). When all available
carbon from this region is consumed by the precipitating carbides, decomposition continues
by the nucleation and growth of σ-phase in the transformed austenitic grain (TA).

*Fe-18Cr-12Mn-0,03C-0,2N steel, quenched from 1150°C and tempered for 30 min at
700°C.* The position of the parent austenite-transformed austenite boundary can be seen due
to the chain of fine carbides (marked by *C*). Note the chromium-rich Cr_2N needles (marked
by *N*), which transform directly into σ-plates.

FIGURE 6.28   The precipitation of σ-phase is accompanied by the migration of the new
high-angle boundary between the transformed austenite and the parent ferrite. It propagates
into ferrite, thus producing a cell of transformed austenite, containing σ-particles.

*Fe-18Cr-12Mn-0,03C-0,2N steel, quenched from 1150°C and tempered for 60 min at
700°C.* The direction of cell propagation is shown by the arrow.

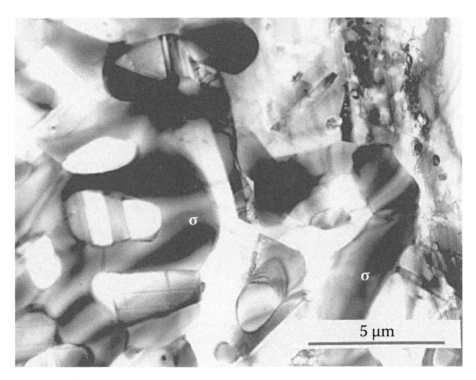

FIGURE 6.29   After a sufficiently long holding period at a higher temperature and with a
sufficiently high amount of chromium present, the σ-phase grows in a skeletal form that can
occupy all the former ferritic grains.

*Fe-18Cr-12Mn-0,03C-0,2N steel, quenched from 1150°C and tempered for 18 h at
700°C.*

6.6 CORROSION RESISTANCE OF PARTICULAR STRUCTURE COMPONENTS OF AUSTENITIC STAINLESS STEELS

TEM investigations of thinned foils that have been exposed to the action of a corrosive media can reveal the behavior of particular components of the microstructure to that media as well as permit to study the initial stages of real corrosion processes. This can provide useful information regarding the role of the microstructure and yield a deeper insight into the processes occurring during real service conditions.

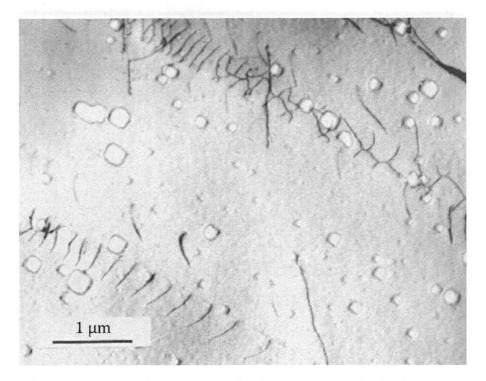

1 µm

FIGURE 6.30 The term *uniform* or *general corrosion* covers all nonlocalized processes of physical-chemical interaction at the interfaces between metals and alloys and the environment. The main feature of this type of corrosion is that it develops uniformly over the whole surface of the material.

Fe-18Cr-12Mn-0,03C-0,6N steel, quenched from 1150°C. The thin foil was exposed for 72 h to the action of 6%FeCl$_3$·6H$_2$O solution at 20°C. The numerous bright spots on the surface of the thin foil are etched by the aggressive media. Note that they develop independently of the existing dislocations. This means that the individual dislocations have not served as initiating sites for corrosion.

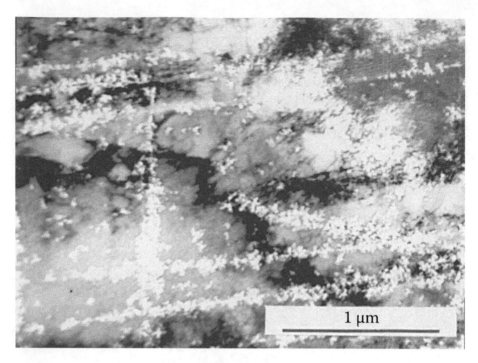

FIGURE 6.31

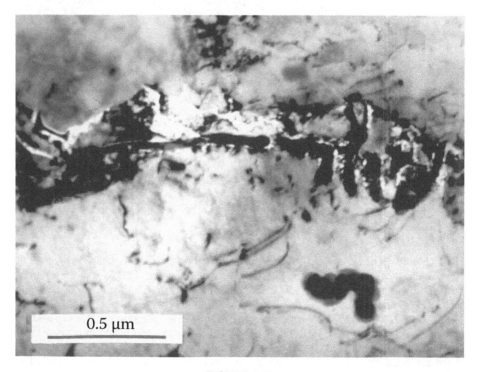

FIGURE 6.32

FIGURE 6.31 *Pitting corrosion* is an extremely dangerous type of electrochemical failure of alloys. This type of corrosion develops in small localized areas *(pits)* where the electrochemical properties of the metal differ from the properties of the surrounding matrix. The enhanced localized anode reaction can be provoked by any chemical or structural discontinuity—nonmetallic inclusions, coarse intermetallic phases, localized defects in the structure, precipitated compounds, and so on. It takes place in metals that rely on passive oxide film for their corrosion resistance (stainless steel falls in this category) in the cases when a break or scratch in the protecting passive film occurs. Regardless of their small size, the pits can be very damaging because they develop in the metal as deep perforations.

Fe-18Cr-12Mn-0,03C-0,6N steel, quenched from 1150°C and 20% cold rolled. The thin foil was exposed for 24 h to the action of 6%FeCl$_3$·6H$_2$O at 20°C. The deformation bands produced by dislocation movement during cold rolling differ in their electrochemical characteristics from the properties of surrounding metal and make the metal in the bands susceptible to the formation of pits. The numerous bright spots along the bands shown in this micrograph are pits etched by the corrosive solution.

FIGURE 6.32 Heating of high-alloyed Cr-Ni austenitic steels at temperatures in the "sensitizing range" from 600°C to 700°C causes precipitation of carbides at the grain boundaries. When the carbon content is higher, carbides can form continuous chains along the boundaries. During their growth, they consume significant amounts of chromium from the surrounding matrix, thus producing chromium-depleted zones. This reduced chromium content lowers the corrosion resistance of the material and makes the regions around the grain boundaries and around the intergranular precipitates especially susceptible to *intercrystal (intergranular) corrosion*. This type of corrosion is a very serious problem not only for the austenitic stainless steels but also for another important group of engineering alloys—Al-Cu alloys are the most popular example.

Fe-18Cr-10Ni steel, quenched from 1150°C and tempered for 60 min at 650°C. The thin foil has been exposed for 12 min to the action of boiling 65% HNO$_3$. The areas around carbides were dissolved by the corrosive media and thus look bright in the micrograph. This indicates the beginning of intergranular corrosion.

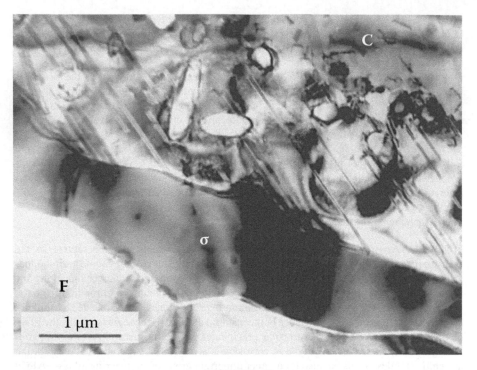

FIGURE 6.33

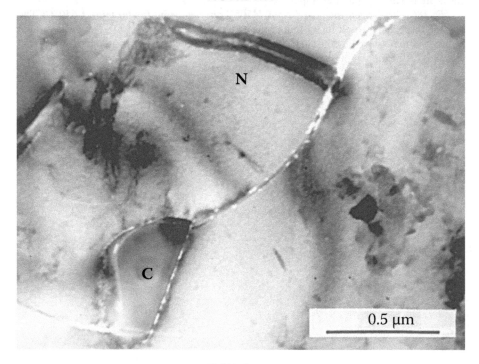

FIGURE 6.34

FIGURE 6.33   The grain boundary precipitation of the chromium-rich σ-intermetallide in the sensitizing temperature range that follows the precipitation of chromium carbides in high-alloy ferrite-austenitic steels leads to enhanced intergranular corrosion, caused again by the chromium-depleted zones around precipitates.

Duplex Fe-18Cr-12Mn-0,02C-0,2N steel, quenched from 1150°C and tempered for 60 min at 700°C. The thin foil has been exposed for 50 min to the action of a boiling solution of $CuSO_4$ in H_2SO_4. The most severe corrosion attack is observed at the σ-phase interface with the ferrite grain into which it grows (marked by *F*) followed by the interfaces of carbides (marked by *C*) with the parent austenite grain. This indicates that the chromium depletion is at a maximum around the σ-phase. This is the most serious factor in the susceptibility of the steel to intergranular corrosion.

FIGURE 6.34   The heat treatment of high-nitrogen austenitic steels at temperatures within the sensitizing region causes discontinuous precipitation of chromium nitrides (see Chapter 5, Section 5.3).

Austenitic Fe-18Cr-12Mn-0,02C-0,45N steel, quenched from 1150°C and tempered for 15 min at 600°C. The thin foil has been exposed for 5 minutes to the action of boiling 65%HNO_3. The number of chromium nitrides on the grain boundary is relatively low due to the short tempering time, but in spite of this, the short period of corrosive attack has been sufficient to affect the whole length of the boundary. This is due to the grain boundary diffusion-controlled mechanism of nitride nucleation, which causes grain boundary depletion of chromium at a significant distance along the boundary and makes the steel susceptible to intergranular corrosion.

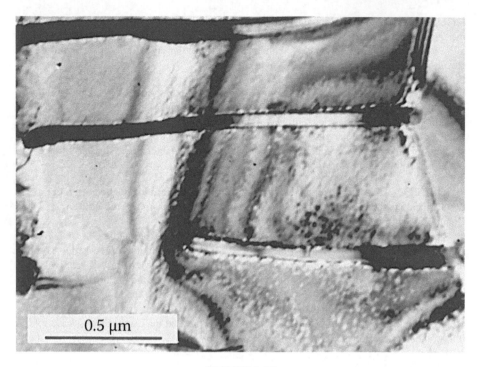

FIGURE 6.35

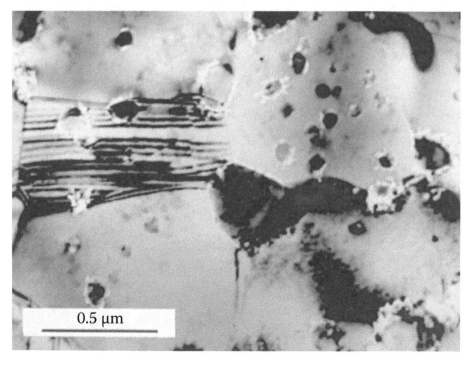

FIGURE 6.36

FIGURE 6.35 A longer tempering duration results in cellular precipitation of lamellar nitride-austenite colonies. The nature of the cellular product (two phases with different composition) and the diffusional mechanism of the cell growth determine the effect of microstructure on the corrosion behavior of the alloy.

Austenitic Fe-18Cr-12Mn-0,02C-0,45N steel, quenched from 1150°C and tempered for 2 h at 600°C. The thin foil has been exposed for 5 min to the action of boiling 65%HNO$_3$. The fine perforations throughout the austenite lamellae in addition to the dissolution at the lateral austenite/nitride interfaces indicate that the whole volume of the transformed austenite is susceptible to the action of the aggressive media. This behavior is usually related to general corrosion, but in the high-nitrogen steels the intergranular corrosion dominates even at the stage of 100% cellular decomposition.

FIGURE 6.36 Cold working applied prior to aging helps to avoid the detrimental effect of discontinuous precipitation with regard to both corrosion and mechanical properties of high-nitrogen austenitic steels by changing the decomposition mechanism from a cellular to a continuous one (see Chapter 6, Section 6.3).

Austenitic Fe-18Cr-12Mn-0,02C-0,45N steel, quenched from 1150°C, cold rolled to a 75% reduction, and tempered for 2 h at 600°C. The thin foil has been exposed for 5 min to the action of boiling 65%HNO$_3$. The areas around all the nitride particles and all the grain boundaries were dissolved by the corrosive media, but their large number and large total length prevented the corrosion from being localized. The process developed as a general corrosion rather than as intergranular corrosion.

6.7 CHARACTERIZATION OF FERRITE IN WELDS OF AUSTENITIC STEELS

The microstructural characterization of welds has two purposes: to evaluate the microstructure with respect to the properties of the material as well as with respect to the welding process used. One of the most important objectives of weld characterization is the quantity and the morphology of the ferrite. Besides strength and ductility, the ferrite distribution influences the susceptibility of the weld to solidification cracking, especially in austenitic stainless steels. The application of TEM characterization is particularly useful in the cases of welds with finely dispersed ferrite and with precipitation on the ferrite-austenite boundaries.

TEM replicas were utilized for the investigation of larger areas of the welds while studies of thin foils permitted a more precise characterization of the microstructure, particularly with reference to precipitation and the accumulation of defects in different regions of the weld.

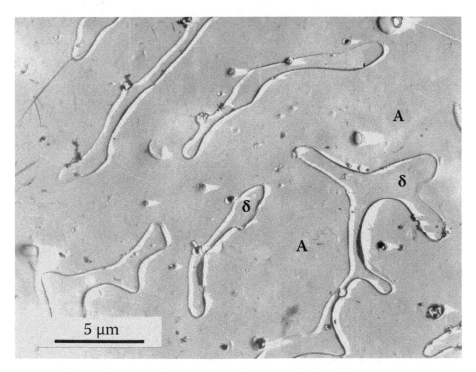

FIGURE 6.37 *Replica of the central part of a weld of austenitic stainless steel Fe-18Cr-9Ni.* The weld was produced by gas-shielded arc welding without filler. The microstructure of the central part of the weld is columnar austenite with δ-ferrite. This morphology of ferrite, which does not reflect its internal crystalline symmetry, is denoted by the term *allotriomorphic.* Allotriomorphic ferrite typically nucleates at the austenite grain surfaces and forms layers at austenite grain boundaries.

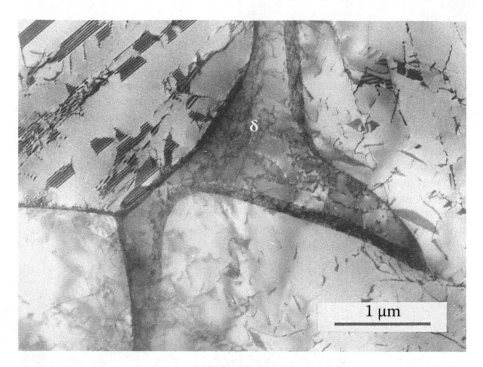

δ

1 µm

FIGURE 6.38

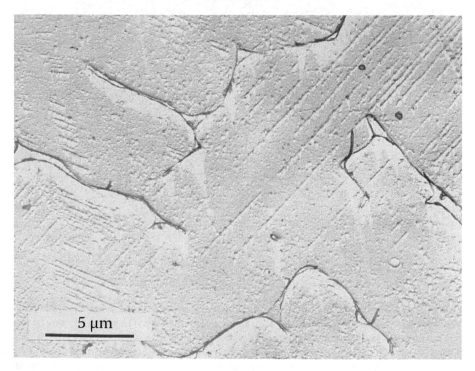

5 µm

FIGURE 6.39

FIGURE 6.38 *A thin foil prepared from the same area as in Figure 6.37.* The grains of aus-tenite can be easily distinguished by the arrow-like arrangement of split dislocations and SFs. The grain of δ-ferrite, which follows the austenite grain contour, contains a three-dimension-ally distributed dislocations.

FIGURE 6.39 *A replica of the same weld, taken from the weld periphery.* The microstruc-ture toward the weld periphery becomes dendritic. The quantity of ferrite diminishes, and the ferrite grains get narrower and form elongated streaks along the solidification direction. The straight lines visible in the interior of the austenite are slip lines aligned along the traces of the fcc close-packed planes.

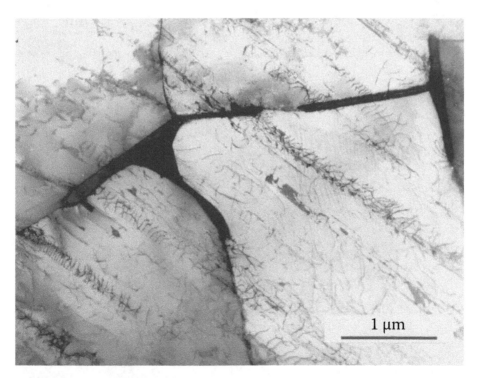

1 μm

FIGURE 6.40 *Examination of thin foils from the periphery of the weld shows that the straight lines that are visible in the previous replica, Figure 6.39, result from a concentrated movement of dislocations in some slip planes of austenite.* The strain arises from the thermal stresses caused by the solidification of the weld. Note that the rows of dislocations pass freely through austenite-austenite boundaries without changing their direction—that is, these are low-angle boundaries.

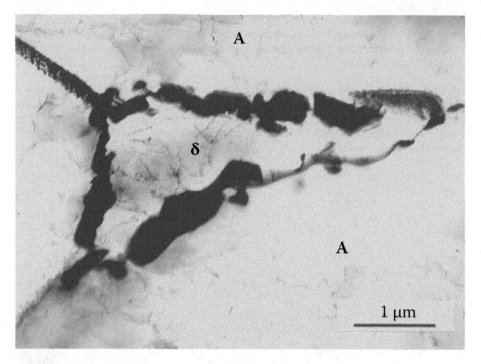

FIGURE 6.41

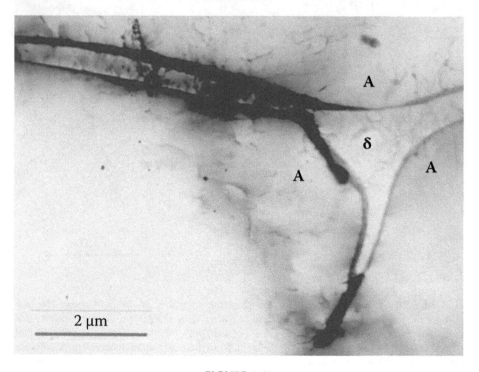

FIGURE 6.42

FIGURE 6.41 *Alloying of the base Fe-18Cr-9Ni steel with some additional elements intended to improve corrosion resistance can provoke phase precipitation during the welding.* For example, the addition of 2.6%Mo causes precipitation of carbides, followed by growth of intermetallic σ-phase on the ferrite-austenite boundaries of the weld. This precipitation sequence is similar to that occurring in the quenched high-chromium ferrite-austenitic steels in the "sensitizing" temperature range (see Chapter 6, Section 6.5). The process is very detrimental to the mechanical properties and the corrosion resistance of the welds, especially when the amount of ferrite present is reasonably high.

FIGURE 6.42 *The additional alloying of the Fe-18Cr-9Ni alloy with low concentrations of nitrogen (in this case, with 0.16%N) prevents the σ-phase formation but causes unacceptable precipitation of chromium nitrides, which cover a significant length of the ferrite-austenite boundaries.* The nitride precipitation runs not only in the weld but also in the heat-affected zone.

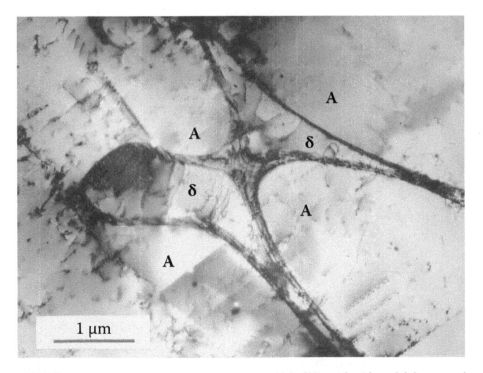

FIGURE 6.43 *The combined alloying of the Fe-18Cr-9Ni steel with molybdenum and nitrogen successfully prevents the precipitation of both carbides and of intermetallic phases in all parts of the weld and in the base metal.*
Weld of stainless austenitic steel containing 2.9%Mo and 0.16%N, gas-shielded arc welding without filler. Weld center. All grain boundaries of the weld are free of precipitates.

Recommended Literature for Further Reading

INTRODUCTION

Fultz, B. and Howe, J.M., *Transmission Electron Microscopy and Diffractometry of Materials,* 2nd ed., Springer-Verlag, Berlin, 2002.

Goodhew, R.J. and Humphreys, E.J., *Electron Microscopy and Analyses,* 2nd ed., Taylor & Francis Group Ltd., London, 1988.

Hirsch, P.B., Howie, A., Nicholson, R.B., Pashley, D., and Whelan, M.J., *Electron Microscopy of Thin Crystals,* Krieger Publishing Co., Melbourne, 1977.

Thomas, G. and Goringe, M.J., *Transmission Electron Microscopy of Materials,* John Wiley & Sons, New York, 1979.

Williams, D.B. and Carter, C.B., *Transmission Electron Microscopy: A Textbook for Materials Science,* Plenum Press, New York, 1996.

CHAPTER 1

Anderson, J.C., Leaver, K.D., Rawlings, R.D., and Alexander, J.M., *Materials Science,* 4th ed., Chapman and Hall, London, 1990.

Barrett, C. and Massalski, T., *Structure of Metals*, 3rd ed., Pergamon Press, 1993.

Callister, W.D., Jr., *Materials Science and Engineering: An Introduction*, 6th ed., John Wiley & Sons, New York, 2003.

Hull, D. and Bacon, D.J., *Introduction to Dislocations,* 4th ed., Butterworth-Heinemann, Oxford, 2001.

Novikov, I.I. and Rozin, K.M., *Kristallografia i Defecti Kristallicheskoj Reshetki Metallov (Crystallography and Defects of Metals Crystal Lattice)*, Moscow, Metallurgia, 1990 (in Russian).

Schtremel, M.A., *Prochnost Splavov. Defecti Reshetki (Strength of Alloys. Lattice Defects)*, Moscow, Metallurgia, 1997 (in Russian).

Smallman, R.E., *Modern Physical Metallurgy,* 4th ed., Butterworths, London, 1985.

CHAPTER 2

Bernstein, M.L., *Struktura Deformirovannih Metallov (Structure of Deformed Metals),* Moscow, Metallurgia, 1977 (in Russian).

Callister, W.D., Jr., *Materials Science and Engineering: An Introduction*, 6th ed., John Wiley & Sons, New York, 2003.

Honeycombe, R.W.K., *The Plastic Deformation of Metals,* Edwards Arnold (Publishers) Ltd., London, 1968.

Smallman, R.E. and Bishop, R.J., *Modern Physical Metallurgy and Materials Engineering,* Butterworth-Heinemann, 1999.

CHAPTER 3

Cahn, R.W., Recovery and recrystallization, in *Physical Metallurgy,* 4th ed., Vol. 3, Cahn, R.W. and Haasen, P., Eds., North-Holland Publishing Co., Amsterdam, 1996, chap. 28.

Cotterill, D. and Mould, P.R., *Recrystallization and Grain Growth in Metals,* Surrey Univ. Press, London, 1976.

Gorelik, S.S., *Rektistallizazia Metallov i Splavov (Recrystallization of Metals and Alloys),* Moscow, Metallurgia, 1978 (in Russian).

Humphreys, F.J. and Hatherly, M., *Recrystallization and Related Annealing Phenomena,* Pergamon Press, Oxford, 2004.

CHAPTER 4

Chalmers, B., *Principles of Solidification,* John Wiley & Sons, New York, 1977.

Flemings, G.A., *Solidification Processing,* McGraw-Hill, New York, 1964.

Porter, D.A. and Easterling, K.E., *Phase Transformations in Metals and Alloys,* 2nd ed., Chapman & Hall, London, 1992.

CHAPTER 5

Bernstein, M.L. and Rachstadt, A.G. (Eds.), *Metalovedenie I Termicheskaia Obrabotka Stali (Physical Metallurgy and Heat Treatment of Steel),* 3rd ed., Vol. 2, Metallurgia, Moscow, 1983, chaps. 25 and 26 (in Russian).

Bhadeshia, H.K.D.H., *Bainite in Steels,* 2nd ed., Institute of Materials, London, 2001.

Chadwick, G.A., *Metallography of Phase Transformations,* Butterworths, London, 1972.

Christian, W.J., *The Theory of Phase Transformations in Metals and Alloys,* 3rd ed., Pergamon Press, Oxford, 2002.

Doherty, R.D., Diffusive phase transformations, in *Physical Metallurgy,* 4th ed., Vol. 2, Cahn, R.W. and Haasen, P., Eds., North-Holland Publishing Co., Amsterdam, 1996, chap. 15.

Hillert, M., *Phase Equilibria, Phase Diagrams and Phase Transformations—Their Thermodynamic Basis,* Cambridge University Press, Cambridge, 1998.

Martin, J.W., *Precipitation Hardening,* Pergamon Press, Oxford, 1968.

Martin, J.W., *Micromechanisms in Particle-Hardened Alloys,* Cambridge University Press, Cambridge, 1980.

Novikov, A.A., *Teoria Termicheskoj Obrabotki Metallov (Theory of Heat Treatment of Metals),* 4th ed., Metallurgia, Moscow, 1986 (in Russian).

Porter, D.A. and Easterling, K.E., *Phase Transformations in Metals and Alloys,* 2nd ed., Chapman & Hall, London, 1992.

Smallman, R.E., *Modern Physical Metallurgy,* Butterworths, London, 1985.

Wayman, C.M. and Bhadeshia, H.K.D.H., Phase transformations, nondiffusive, in *Physical Metallurgy,* 4th ed., Vol. 2, Cahn, R.W. and Haasen, P., Eds., North-Holland Publishing Co., Amsterdam, 1996, chap. 16.

Index

A

Aging, 80
Allotriomorphic ferrite, 163–167
Amorphous metals and alloys, 73–75, *see* Glassy
    metals; Metallic glasses

B

Bainite, 121
    lower, 122–123
    upper, 124–125
Bainite plates, *see* Bainitic laths
Bainite subunits, *see* Bainitic laths
Bainitic ferrite, 121
Bainitic laths, 122–125, *see* Bainite plates;
    Bainite subunits
Bainitic transformations, 121
Boundaries, 19
    grain, *see* Grain boundaries
    intrephase, *see* Interfaces
    subgrain, *see* Subgrain boundaries
Burgers vector, 1, 5, 21–22, 40–41

C

Cells, 35, 40, 97
    of discontinuous precipitation, 97, 100–101,
        see Cellular colonies
    of dislocation substructure, *see* Dislocation
        configurations, cells
Cellular colonies, *see* Cells, in discontinuous
    precipitation,
Cellular precipitation, *see* Phase precipitation,
    discontinuous
Chromium depletion, 151, 156–159
Climb, *see* Dislocation movement,
    non-conservative
Clusters, 80
Colonies, eutectoid, *see* Eutectoid, colonies
Continuous precipitation, 80, 81–87
Corrosion, 155
    general, 155, 160–161, *see* Uniform corrosion
    intercrystal, 156–161, *see* Intergranular
        corrosion
    pitting, 156–157
Cross-slip, *see* Dislocation movement,
    non-conservative
Crystal growth, 69

Crystal lattice, 1
    body centered cubic, 29, 30, 34–35
    body centered tetragonal, 109
    close packed hexagonal, 16–17
    face-centered cubic, 34–35
Crystal nucleation, *see* Nucleation, of, crystals
Crystalographic orientation, 87, 103, 109

D

Deformation bands, *see* Slip lines
Deformation contrast, *see* Strain contrast
Deformation twinning, 28–29, 46–47,
    see Mechanical twinning
Dendritic cells in Al-Si alloys, 148–150
Dendritic cells in welds, 163–165
Diffraction, xi
    patterns, 74–77
    rings, 74–77
    spots 74–77, 81– 82
Diffraction contrast, xi
Discontinuous precipitation, 97–101, 158–161,
    see Cellular precipitation
Dislocation configurations, 31
    cells, 35–37, 44–47
    complexes, 20–21, 31, 34–35, 39, *see*
        Dislocation configurations, tangles
    loops, 16–17, 95, 146
    pileups, 12–13, 22–23
    subboundaries, 20–21, 40–41; *See also*
        Subgrains
    tangles, *see* Dislocation configurations,
        complexes
    walls 20–21, 35
Dislocation density, 1, 4–5, 11, 34–37, 141
Dislocation distribution, 1, 31
    three-dimensional, 4–7, 33–35, 43, 164–165
    two-dimensional, 7–10, 49–53, 128–129, 151;
        See also Planar dislocation
        distribution
Dislocation lines, 4–5, 94–95
Dislocation martensite, *see* Martensite, lath
Dislocation movement, 1
    conservative, 12–13, 33, 49, *see* Glide; Slip
    non-conservative, 6–7, 39, 57, 130–131,
        see Climb; Cross-slip
    obstacles for, 10, 11, 22–23, 94–95
Dislocation initiation, 11–13, 22–23
Dislocation interaction, 1, 20–21, 34–35
Dislocation multiplication, 11–13

171

Dislocation pinning, 94–95
Dislocation splitting, 7–10, 49–51
Dislocation substructure, 1, 31–33, 40–41
    cell, 35–37
    grid, 53, 141
Dislocations, 1
    extended, 7–10, 49–51, 128–129, *see* Split
        dislocations
    helical, 5–7, 88–89
    partial, 7–10, 49–51, 128–129, *see* Imperfect
        dislocations
    screw, 69–70
    sessile, *see* Lomer-Cottrell barrier
    unit, 4–5, 43, 57–58, *see* Perfect dislocations

E

Electron diffraction, 1
Eutectic, in Al-Si alloys 76–77,148–150
Eutectoid decomposition, *see* Eutectoid
        transformation,
Eutectoid, 103
    colonies, 103–107, *see* Eutectoid grains
    grains, *see* Eutectoid colonies
Eutectoid transformation, 103–107, *see* Eutectoid
        decomposition

F

Faulted martensite, *see* Martensite, plate
Feathery bainite, 124–125; *See also* Bainite, upper
δ Ferrite, in welds, 163–167
Frank-Read sourses, 11–13, 22–23

G

Glassy metals, *see* Amorphous metals
Glide, *see* Dislocation movement, conservative
Gold decoration, 69–71
GPZ, *see* Guinier-Preston zones
Grain boundaries, 19
    high-angle, 21–23
    low-angle, 165
Grain boundary processes, 64, 89
    migration, 64–67, 90–91, 142–143, 152–153
    pinning, 90–91
    precipitation, 89–91, 152–153, 166–167
    segregation, 89–91
    slip, 141, 145
Grain growth, 59, 64–65, 143–145
Grains, 59, 69
    accommodation, 142–143, 145–146
    coarsening, 59, 66–67
    elongation, 143–145

    nucleation, *see* Nucleation of grains
    refinement, 147; *See also* Modification
Growth ledges, *see* Growth steps
Growth spirals, 70–71
Growth steps, 60–71, *see* Growth ledges
Guinier-Preston zones, 80–83, 87,94–95,
        see GPZ; Zone precipitation

H

Habit plane, 81–87, 99, 107, 117, 123
Heat treatment, of Al Si alloys, 149
Heterogenous distribution of precipitates,
        see Precipitations distribution,
        localized
Homogenous distribution of precipitates,
        see Precipitations distribution,
        uniform

I

Image contrast, xi
Imperfect dislocations, *see* Dislocations, partial
Interfaces, 90, 93, *see* Interphase boundaries,
    austenite-ferrite, 151–153, 167
    matrix-precipitate, 84–85
Interference fringes, xii, 8–9, 21–23, 26–27,
        44–45, 49
Intergranular corrosion, *see* Corrosion, intercrystal
Interlamellar spacing, 79, 97, 103
    in cellular colonies, 97, 98–101
    in eutectoid colonies, 103–105
Intermediate phases, *see* Phases, metastable

L

Lattice defects, *see* Lattice imperfections
Lattice imperfections, 1, *see* Lattice defects
Lomer-Cottrell barrier, 10, *see* Dislocations,
        sessile; Stair-rod dislocations

M

Massive martensite, *see* Martensite, lath
Martensite, 109
    lath, 114–115, 116–117, *see* Massive
        martensite; Dislocation martensite
    α-martensite, 116–117
    ε-martensite, 50–51, 116–117, *see* Sheet
        martensite
    needle-like, 112–113
    plate, 110–113, 118–119, *see* Twinned
        martensite; Faulted martensite
    strain-induced, 113, 116–117

tempered, *see* Tempered martensite, 117,
    122–123
  units, 111, 114–115
Martensitic transformations, 109
Mechanical twinning, *see* Deformation twinning
Metallic glasses, *see* Amorphous metals
Metastable phases, *see* Phases, metastable
Microcrystalline alloys, 76–77; *See also* Rapidly
    solidified microcrystalline alloys
Microtwins; 29–30; *See also* Twins, deformation
Midrib, 110
Misfit, 82
  contrast, 82–85
  dislocations, 84–85
Misorientation angle
  between cells, 40–41, 57–58
  between grains, 19, 100–101
  between subgrains, 20–21
Modification of cast microstructure, 147, 150

N

Nucleation, of
  crystals, 73–75, *see* Crystal nucleation
  phases, 79; *See also* Phase precipitation
    heterogenous, 86–91, 98–99
    homogenous, 81–87

O

Orientation
  preferred, 122–123
  relationship, 86–87, 103, 109

P

Packed martensite, *see* Lath martensite
Pearlite colonies, 104–107; *See also* Eutectoid
    colonies
PFZ, *see* Precipitation-free zones
ε-Phase, *see* Martensite, ε-martensite
σ-Phase, 151–153, 166–167
Phase precipitation, 79, 133
  continuous, 80–87, 160–161
  discontinuous, 97–101, 138–139, 158–161,
    see Cellular precipitation
Phases, 79
  coherent, 82–85
  equilibrium, 84–85, 98–99, 131–132
  metastable, 80, 82–83, *see* Intermediate
    phases; Transition phases
  semicoherent, 84–85
Phase transformations, 79
Planar dislocation distribution, *see* Dislocation
    distribution, two-dimensional

Plastic deformation,
  cold, 34–37, 43–53, 160–161
  hot, 57–58
  warm, 55–57
Point defects, *see* Vacancies
Polygonisation, 59
  dynamic, 66–67
  static, 62–65
Precipitation-free zones, 90–91, 134–137, *see* PFZ
Precipitations, distribution, 80
  uniform, *see* Homogenous distribution of
    precipitates, 80–85, 89, 135, 138–39,
    142–143, 148–149
  localized, *see* Heterogenous distribution
    of precipitates, 86–87, 89–91, 129,
    131–137, 152–153
Precipitation strengthening, 81–85, 133

R

Rapidly solidified microcrystalline alloys, 69,
    76–77
Rapid solidification process, 73, 76–77
Recovery, 59, 62–63, 66–67
Recrystallization, 59
  dynamic, 68, 130–131
  primary, 62–65
Recrystallization centers, 62–65, 142–143
Recrystallization twins, *see* Twins, annealing
Replica, 1, 106–107, 124–125, 148
Resolving power of TEM, 1
Retained austenite, 124–125

S

Sensitization of stainless steels, 157–161
SF, *see* Stacking faults
SFE, *see* Stacking faults energy
Sheet martensite, *see* ε-Martenste
Slip, *see* Dislovation movement, conservative
Slip bands, *see* Slip lines
Slip lines, 46–47, 51–53, 156–157, *see* Slip bands;
    Deformation bands
Solute depletion, 88–89, 90–91; *See also*
    Chromium depletion
Solute segregation, 87; *See also* Grain boundary,
    segregation
Softening mechanisms, 59
Spiral growth, 70–71
Split dislocations, *see* Disliocations, extended
Stair-rod dislocations, *see* Dislocations, sessile
Stacking fault energy, 1, 8–9, 28–29, 31, 49, 55,
    see SFE
Stacking faults, 1, *see* SF
  ribbons, 8–9, 49–51, 151–153, 164–165
  width, 8–9, 43, 49, 52–53

Strain contrast, xi, 22–23, 81–85, *see* Deformation
        contrast
Strain fields, 94–95
Strain hardening, 59, 93–95, *see* Work hardening
Subboundaries, 20–21, 41–42, 62–63, 130–131,
        see Boundaries, subgrain
Subgrains, 20–21, 40–41, 62–63
Superplasticity, 141

T

Tempered martensite, *see* Martensite, tempered
Thin foils, 1
Three-dimensional dislocation distribution,
        see Dislocation distribution,
        three-dimensional
Transition phases, *see* Phases, metastable
Triple dislocation nodes, 9
Twin boundaries, 25, 28–29
        coherent, 25–27
        nonncoherent, 25– 27
Twinned martensite, *see* Marteniste, plate
Twins, 25
        annealing, 27, 66–67, *see* Recrystallization
                twins
        deformation, 29, 30, 44–47, 50–51; *See also*
                Microtwins
Two-dimensional dislocation distribution,
        see Dislocation distribution, planar

U

Ultra-fine grains, 76–77, 138–139, 142–143,
        144–146
Uniform corrosion, *see* Corrosion, general

V

Vacancies, xi, 5, 15–18, 80, 88–89, *see* Point
        defects
        depleted zones, of, 90–91
        sinks, for,15, 17, 90–91
        sources, of, 15, 17
        supersaturation, 15, 17, 80, 89
Vacancy configurations, xii
        complexes, 15–18, 88–89
        loops, 15–18, 143, 146

W

Widmanstätten morphology, 86–87, 107
Work hardening, *see* Strain hardening

Z

Zone precipitation, *see* Guinier-Preston zones